Ordered Porous Nanostructures
and Applications

Nanostructure Science and Technology

Series Editor: David J. Lockwood, FRSC
National Research Council of Canada
Ottawa, Ontario, Canada

A Continuation Order Plan is available for this series. A continuation order will bring delivery of each new volume immediately upon publication. Volumes are billed only upon actual shipment. For further information please contact the publisher.

Ordered Porous Nanostructures and Applications

Edited by

Ralf B. Wehrspohn

Department of Physics
University of Paderborn
Paderborn, Germany

 Springer

Library of Congress Cataloging-in-Publication Data

Wehrspohn, Ralf B.
    Ordered porous nanostructures and applications / Ed. by Ralf B. Wehrspohn.
        p. cm.—(Nanostructure science and technology)
    Includes bibliographical references and index.
    ISBN 0-387-23541-8
    1. Nanotechnology.   2. Nanostructures.   I. Title.   II. Series.

    T174.7.W44 2005
    620′.5—dc22   2004062627

ISBN 0-387-23541-8

Printed in the United States of America.

9 8 7 6 5 4 3 2 1

springeronline.com

Contributors

J. Carstensen
Material Science Department,
Faculty of Engineering
Christian-Albrechts University,
Kaiserstraße 2, D-24143, Kiel,
Germany

J.-N. Chazalviel
Laboratoire de Physique
   de la Matière Condensée,
CNRS-Ecole Polytechnique,
91128 Palaiseau Cedex,
France

M. Christophersen
Material Science Department,
Faculty of Engineering
Christian-Albrechts University,
Kaiserstraße 2, D-24143, Kiel,
Germany

H. Föll
Material Science Department,
Faculty of Engineering
Christian-Albrechts University,
Kaiserstraße 2, D-24143, Kiel,
Germany

P.J. French
Electronic Instrumentation Laboratory,
Department of Microelectronics,
Faculty of Electrical Engineering,
Mathematics and Computer Science,
Delf University of Technology,
Mekelweg 4, 2628 CD Delf,
The Netherlands

L.V. Govor
Institute of Physics,
University of Oldenburg,
D-26111 Oldenburg,
Germany

Siegmund Greulich-Weber
Physics Department,
Faculty of Science,
University of Paderborn,
D-33095 Paderborn,
Germany

Riccardo Hertel
Dept. of Solid State
Research, Research Center Juelich,
D-52425 Juelich,
Germany

S. Langa
Material Science Department,
Faculty of Engineering
Christian-Albrechts University,
Kaiserstraße 2, D-24143, Kiel,
Germany
and
Laboratory of Low Dimensional
   Semiconductor Structures,

Technical University of Moldova,
St. cel Mare 168, MD-2004, Chisinau,
Moldova

V. Lehmann
Infineon Technologies AG,
Dept. CPS EB BS,
Otto-Hahn-Ring 6,
D-81730 München,
Germany

Heinrich Marsmann
Faculty of Science,
University of Paderborn,
D-33095 Paderborn,
Germany

Hideki Masuda
Department of Applied Chemistry,
Tokyo Metropolitan University,
1-1 Minamiosawa, Hachioji,
Tokyo 192-03,
Japan

Kornelius Nielsch
Max-Planck-Institute of
    Microstructure Physics,
Weinberg 2D-06120 Halle,
Germany

H. Ohji
Mitsubishi Electric Corporation,
Advanced Technology Research
    and Development Centre,
Amagasaki, Hyogo 6618661,
Japan

F. Ozanam
Laboratoire de Physique
    de la Matière Condensée,
CNRS-Ecole Polytechnique,
91128 Palaiseau Cedex,
France

Joerg Schilling
California Institute of Technology,
Pasadena, CA 91125,
USA

I.M. Tiginyanu
Laboratory of Low Dimensional
    Semiconductor Structures,
Technical University of Moldova,
St. cel Mare 168, MD-2004, Chisinau,
Moldova

Ralf B. Wehrspohn
Department of Physics,
University of Paderborn,
D-33095 Paderborn,
Germany

Foreword

Numerous major advances in research and technology over the last decade or two have been made possible by the successful development of nanostructures made of metals, insulators and especially semiconductors. Nanostructures are man-made objects that have one, two or three dimensions in the sub-micrometre to nanometre regime. Nanostructures made of semiconductor quantum wells, which consist of alternating layers of two different semiconductors with typical thicknesses in the sub-10 nm regime, were first demonstrated more than 20 years ago. Today, they are at the heart of most semiconductor lasers. More recently, carbon nanotubes and semiconductor quantum dots have attracted a lot of scientific attention because of their unique properties and their wide-ranging potential applications. Even the dominant industry since the late-20th century have embraced the use of nanostructures. Indeed, in the microelectronic industry, the size of individual transistors is well below 100 nm and within 10 years may approach the regime where quantum size effects start playing a role.

One significant difficulty with nanostructures is how to prepare them. One can distinguish two approaches: top-down and bottom-up. In the top-down approach, objects of ever-smaller dimensions are carved out of larger objects. This approach is taken in the semiconductor industry where advanced lithography aided by specific steps such as selective oxidation has unrelentlessly shrunk the typical dimensions to well below 1 μm. However, this approach is increasingly complicated and expensive. The bottom-up approach consists of growing small objects to their desired size and shape. This is usually accomplished by chemical means. This approach is very flexible and usually inexpensive, but it too suffers from significant problems, chief among them are size and positioning control and throughput.

Porous nanostructures have attracted a lot of attention because they combine many of the advantages of the top-down and bottom-up approaches. The typical dimension can be varied from a few nanometres to many micrometres, the porous structures be made in many materials and be ordered, and entire wafers can be processed in minutes. Since 1990, a lot of effort has been devoted to understanding and controlling the pore formation mechanism and to evaluating the usefulness of porous nanostructures in technology. This book, edited by Ralf Wehrspohn, is a very timely and excellent review of the state of the art in ordered porous nanostructures and their applications. It contains nine chapters written by leading experts. The chapters on materials and preparations cover the most important porous materials, namely silicon, III–V semiconductors, alumina and polymers. These chapters cover all the important aspects of the fascinating materials science of porous materials. Topics ranging from well-understood phenomena to still controversial observations are discussed. The second part of the book is devoted to applications. The last three chapters cover the important applications in optics, magnetics and micromachining.

This book will be valuable to all researchers active in the field, whether they are experienced or just starting, and whether they are in research or development.

Philippe M. Fauchet
Rochester, NY

Preface

In the 1990s, a variety of two-dimensional self-ordered porous nanostructures were discovered. Starting with ordered macroporous silicon discovered by Lehmann and Föll in 1990, other self-ordered materials were discovered: self-ordered porous alumina by Masuda and Fukuda in 1995, self-ordered diblock copolymers aligned on substrates by the Russel group in 1994, self-ordered zeolites (MCM-41) by the Mobil Oil group in 1992, self-ordered porous polymer structures with honeycomb morphology by Francois and co-workers in 1994 and finally self-ordered porous group III–V semiconductors by Föll and co-workers in 1999. Similarly, also three-dimensional self-ordered nanostructures developed in the same decade like three-dimensionally arranged block copolymers and 3D colloidal self-assembly.

This edited book presents the synthesis of the five materials systems mentioned above and tries to explain the physical and chemical mechanisms of self-ordering. In general, ordering is always due to repulsive or attractive forces between the pores leading in two dimensions to the hexagonal lattice. In three dimensions, stacking can either lead to the fcc or hcp lattice, but it is always a closed-packed configuration. These ordered porous nanostructures are very attractive for template synthesis of nanowires or nanotubes or in 3D even of more complex structures, and a number of examples of ordered porous nanostructures are given in different chapters.

The last three chapters describe three very prominent areas of applications of these materials: photonics, magnetic storage media and nano-electromechanical systems (NEMS).

Ralf Wehrspohn
Paderborn, Germany

Contents

I

MATERIALS AND PREPARATIONS

1

Electrochemical Pore Array Fabrication on n-Type Silicon Electrodes

V. Lehmann

Infineon Technologies AG, Dept. CPS EB BS, Otto-Hahn-Ring 6, D-81730 München, Germany
volker.lehmann@infineon.com

1.1. WHY THE FIRST ARTIFICIAL PORE ARRAYS WERE REALIZED IN N-TYPE SILICON ELECTRODES

The surface morphology of a solid-state electrode after an electrochemical dissolution process depends sensitively on the parameters of anodization. The one extreme case is an anodization condition under which even a rough electrode surface becomes homogeneously smooth, this process has been termed electropolishing. If in contrast the surface becomes rougher, we deal with corrosion or pore formation. In the prior case, commonly the crevice geometry is random and no narrow size regime is observed. This is different in the latter case. Electrochemically formed porous materials usually show a narrow pore size distribution and a certain pore density, which allows us to determine the ratio of pore volume to the total volume, the porosity. In most cases, the pore distribution at the electrode surface is random; however, in certain cases, like for example, for porous alumina, a short-range order may be observed. Pore arrays with a long-range order of the pore positions, which are desirable for a multitude of applications, can only be produced artificially.

The fact that n-type silicon was the first electrode material on which pore arrays of such a long-range order have been realized is not purely accidental. For artificial patterning, the electrochemical pore initiation process must be understood and a structuring technique must be available. In many electrode–electrolyte systems, the pore initiation mechanism is complex and still under debate, as for example, for pore formation in aluminium [1]. In other systems, like alumina, the pore size is small and just becomes assessable for today's most advanced structuring technologies [2]. In contrast, in low-doped n-type silicon electrodes, to a large extent, the pore initiation is simply controlled by the interface topography and the pore size is well measured in the micrometre regime. This enabled us about a decade ago to use the standard silicon process technology available at that time, such as thin film deposition, photolithography, and wet etching, to generate a pore initiation pattern. Upon anodization in hydrofluoric acid (HF) this arbitrary pattern developed into an array of straight pores [3]. The resulting pore morphology can be inspected by cleaving the silicon electrode and subsequent optical microscopy (OM) or scanning electron microscopy (SEM).

Over the years, the electrochemical etching process has been optimized and today the pore array may cover a whole silicon wafer and penetrate its full thickness. Furthermore, pore arrays have been realized by the anodization of a multitude of other materials as discussed in subsequent chapters.

1.2. THE PHYSICS OF PORE INITIATION ON SILICON ELECTRODES IN HF

In order to predetermine the position of an electrochemically formed pore a detailed understanding of the pore initiation process is required. The pore initiation may, for example, be dominated by the impurity or defect distribution, by the state of electrode passivation, or by the topography of the electrode surface. As a consequence, structuring techniques for electrochemical pore array fabrication could be based on local impurity implantation, depassivation, or etching of depressions to generate pore initiation sites.

For the case of n-type silicon an anodic oxide has been discussed as a potential candidate for passive film formation. An indispensable constituent of all electrolytes for pore formation in silicon electrodes, however, is hydrofluoric acid, which readily dissolves SiO_2. Pore initiation by defects is unlikely as well, because today's silicon crystals are manufactured free of defects and with atomic impurity levels down to 10^{12} cm^{-3}. The third of the above options, the topography of the electrode surface, has been found to be the relevant factor for the pore initiation process.

All pore formation on silicon electrodes is observed in an anodic regime where the dissolution reaction is limited by charge supply from the electrode. This regime is characterized by an anodic current density below a critical value J_{PS}. J_{PS} depends on electrolyte concentration, temperature and crystal orientation of the electrode [4]. The dissolution reaction is initiated by a hole (defect electron), reaching the silicon–electrolyte interface. Since charge supply is the limiting factor dissolution occurs preferentially at sites which attract holes. If no such sites are present, like for example, for an atomically flat electrode the dissolution starts homogeneously. However, any inhomogeneity of dissolution will be amplified and within seconds etch pits are formed, which act as initiation sites for pore growth. This random pore initiation process at polished electrode surfaces is shown in Figure 1.1. Note that in the first 30 seconds of anodization a very high density of nanometre-sized etch pits are generated. In the next 30 seconds, a few of these pits increase their size by a factor of 5 or more consuming neighbouring pits. This process continues until the number of surviving pores becomes constant. This is the case, after about 240 seconds, for an n-type doping concentration of 10^{16} cm^{-3}.

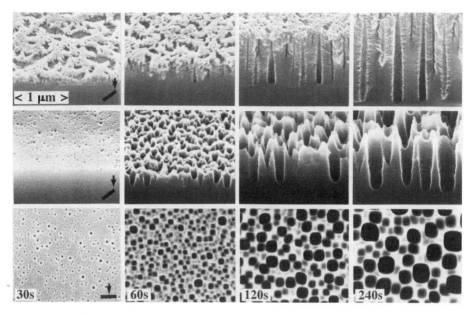

FIGURE 1.1. SEM micrographs of spontaneous pore initiation on polished surfaces of n-type Si electrodes anodized for the indicated times under white light illumination of the front side (14 V, 2.5% HF, 10 mA/cm^{-2}, n-type Si 10^{16} cm^{-3} (100)). Microporous silicon covering the macropores, as shown on the top row (cross-sectional view), has been removed by alkaline etching for better visibility at centre (cross-sectional view) and the bottom row (surface view). From Ref. [1].

If depressions are already present in the electrode prior to anodization, it is easily understood that they become initiation sites for the pore formation. A pattern of artificial initiation sites works only as desired, if its pitch is close to the pore spacing that develops spontaneously on the silicon electrode upon anodization. The average distance of random pores is usually in the same order of magnitude as the pore diameter. The observed diameters of pores formed in silicon electrodes cover four orders of magnitude and is classified in three size regimes. A porous film is designated microporous if the pore diameter is below 2 nm. In this size regime, pore formation is dominated by quantum size effects. While the pore size becomes mesoporous (2 nm < pore diameter < 50 nm) or macroporous (pore diameter > 50 nm) if the formation process is dominated by the electric field in the space–charge region (SCR). In the electric field dominated case, the morphology of the porous structure depends sensitively on the way the charge carriers pass through the SCR. An overview of the different size regimes and the proposed pore formation mechanism is displayed in Figure 1.2. The fact that the pore initiation site can be predetermined by a depression in the electrode surface has first been shown for macropores in low-doped n-type electrodes for which the pore formation is dominated by minority carrier collection. Later on, however, it has been shown that for the other three field-dominated pore formation effects, as displayed in Figure 1.2, a depression is as well sufficient as initiation site [5]. Figures 1.3b–d show arrays of macropores on p-type silicon, the pores initiate preferentially at the pyramidal depression in the pre-structured electrode surface. Figure 1.3e shows a minute mesopore formed by tunnelling of charge

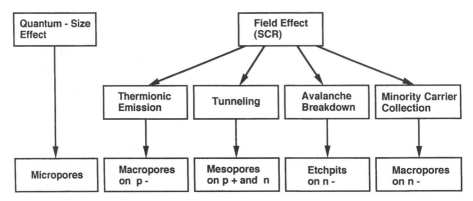

FIGURE 1.2. Effects proposed to be responsible for pore wall passivation (top row). Effects which can lead to passivation breakdown at the pore tip (middle row) and the resulting kind of porous silicon structure together with substrate doping type (bottom row). From Ref. [5].

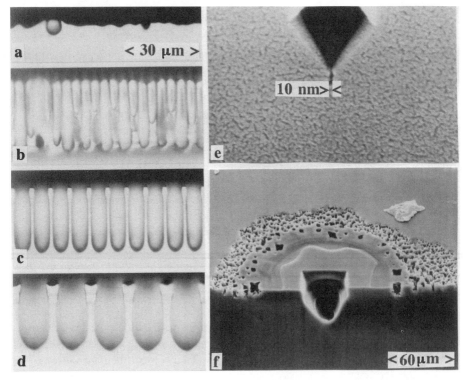

FIGURE 1.3. Electrochemical pore formation in silicon electrodes of different kinds and density of doping initiated by an artificial depression. (a–d) Pore initiation on a polished (a) and on patterned (b–d) p-type silicon electrodes (2 mA/cm^2, 3% HF, 240 minutes, p-type Si 3 × 10^{14} cm^{-3}) (OM after [6]). (e) Mesopore formation at the tip of a pyramidal etch pit (10 V, 6% HF, 5 seconds, n-type Si 10^{15} cm^{-3}) (SEM after [4]). (f) A large circular etch pit structure formed by avalanche breakdown (50 V, 6% HF, 100 seconds, n-type Si 10^{15} cm^{-3}). Note that the structure is centred around the pyramidal initiation etch pit (SEM from Ref. [5]).

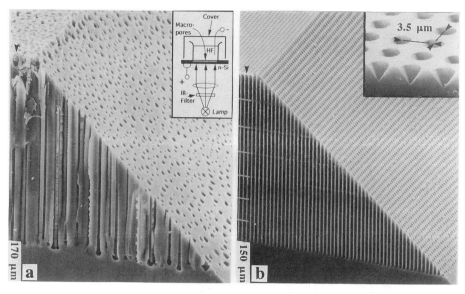

FIGURE 1.4. SEM micrographs of surface, cross section and a 45° level of anodized n-type silicon samples (10^{15} cm^{-3}). Sample (a) shows randomly distributed pores due to anodization of a polished electrode, while sample (b) shows a square array of pores generated by anodization of a patterned surface. The pore initiation pattern, as shown in the inset, has been produced by photolithography and alkaline etching. From Ref. [4].

carriers located at the tip of a depression. Figure 1.3f shows that a large cavity formed by avalanche breakdown is also centred at the tip of an artificial depression. It can be speculated that even the initiation of micropores is sensitive to the electrode topology. In order to test this hypothesis, however, the required resolution of the initiation pattern has to be in the order of 1 nm and is therefore beyond today's photolithographic structuring techniques.

The density of random pores in n-type silicon electrodes decreases with decreasing substrate-doping density. For an n-type doping concentration of 10^{15} cm^{-3}, as shown in Figure 1.4a, the pore initiation process takes longer and the final pore density is lower, as for example shown for an n-type doping concentration of 10^{16} cm^{-3} in Figure 1.1. This dependence of pore density on the doping level of the bulk silicon reflects the influence of the SCR width on the formation process of initiation sites. A depression produces a deformation of the SCR and the electric field becomes maximum where the radius of curvature of the depression has its minimum. For the case of highly doped silicon the electric field easily reaches its breakdown value even for moderate applied bias if the tip radius is reduced to a few tens of nanometres. As a consequence, the tunnelling of charge carriers is confined to the tip of the depression. In low-doped silicon, where the field strength is usually below the breakdown value, the transfer of charge carriers is still influenced by the topography and shows a maximum at the tip of the depression. Even if the field of the SCR is neglected and pure hole diffusion is considered, a depression is still a favourable location for charge transfer. In the case of low-doped n-type silicon electrodes the electric field as well as the diffusion have to be considered for pore initiation and pore growth [5,6].

In conclusion, a flat silicon electrode anodized in HF below the critical current density is unstable. Such a system shows a tendency to enhance inhomogeneities of the surface topography. An artificial pore initiation pattern realized by depression in the electrodes surface exploits this instability to form pore arrays. An example of such a pore array is shown in Figure 1.4b. The array of etch pits used for initiation is shown in the inset of this figure.

1.3. THE PHOTOLITHOGRAPHIC PRE-STRUCTURING PROCESS AND THE ANODIZATION SET-UP

A depression, sufficient for pore initiation in an n-type electrode, can be realized in many ways. Most compatible with today's semiconductor manufacturing techniques is photolithographic structuring. The basic process sequence is sketched in Figure 1.5. A thermal oxide is formed on a polished, (100)-oriented, n-type silicon wafer with a highly doped n-type backside layer. A photoresist is then deposited on the front side and illuminated using a mask with the desired pore pattern. Subsequently windows in the oxide film are opened, using a plasma etch process, for example. A wet alkaline etching

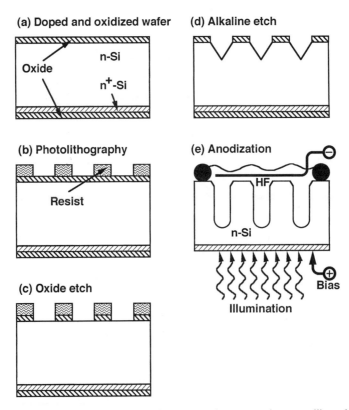

FIGURE 1.5. Schematic view of the fabrication process of pore arrays in n-type silicon electrodes.

process then generates sharp-tipped etch pits that show the geometry of an inverted pyramid, as shown in the inset of Figure 1.4b. These etch pits act as initiation sites for the electrochemical etching process, because their collection efficiency for holes (defect electrons), generated by illumination of the electrode backside, is much better than the one for the flat substrate.

The pore initiation process is sensitive to the geometry of the depression. For a depression with a large radius of curvature at the bottom, for example, the starting position of the pore is badly defined. As a consequence, a certain mispositioning of the pores in an array can be expected. An inverted pyramid with a flat bottom, for example, can be realized by a reduction of alkaline etching time. For such a geometry, the formation of four pores located at the four corners of the inverted pyramid bottom has been observed.

The electrolyte used for macropore array formation in n-type Si electrodes is composed of aqueous HF. The pore growth rate depends sensitively on HF concentration (commonly 1–10%) and is in the order of 1 μm/min. Hydrogen is a by-product of the electrochemical dissolution process. In order to reduce the sticking probability of hydrogen bubbles to the electrode surface, addition of a detergent and strong electrolyte agitation are recommended. PVC or PP is recommended as materials for the cell body. Standard O rings (nitrile polymer) are found to be stable in the HF electrolyte. A platinum electrode is commonly used as cathode. A reference electrode is not required because the pore formation process on n-type silicon is not very sensitive to bias. The highly doped backside electrode is connected to the positive side of the power supply, collecting the photo-generated electrons.

The light source used for illumination of the n-type electrode should emit at wavelengths below 900 nm, because longer wavelengths penetrate deep into the bulk and might generate charge carriers in the pore walls. Such light sources can be realized by LEDs or by a filament lamp with an optical short-pass filter.

1.4. LIMITING FACTORS AND DESIGN RULES FOR MACROPORE ARRAYS ON N-TYPE SILICON ELECTRODES

Not all desirable macropore array geometries can be realized by the electrochemical etching process. This section gives the upper and lower limits for pore dimensions and a few design rules [7].

The realization of a desired pore pattern requires a certain doping density of the n-type Si electrode. A good rule of thumb for the selection of an appropriate substrate is to multiply the desired pore density given (in μm^2) by 10^{16} and take this number as doping density (in cm^{-3}). This dependency is shown in Figure 1.6. A square pattern of 10 μm pitch, for example, produces a pore density of 0.01 pores/μm^2 which can be best etched using a substrate of an n-type doping density of 10^{14} cm^{-3}. A maladjustment of pore pattern and substrate doping density will lead to branching of pores, as shown in Figure 1.7a or dying of pores, as shown in Figure 1.7c.

The number of possible arrangements of the pore pattern is only limited by the requirement that under homogeneous backside illumination the porosity has to be constant on a length scale above about three times the pitch. This means it is possible to etch a

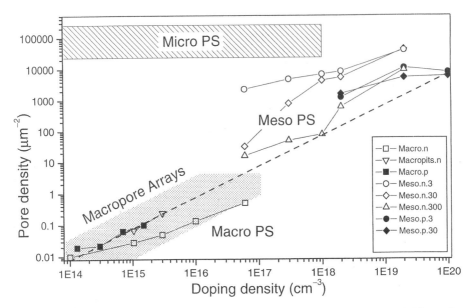

FIGURE 1.6. Pore density versus silicon electrode doping density for porous silicon layers of different size regimes. The dashed line shows the pore density of a triangular pore pattern with a pore pitch equal to two times the SCR width for 3 V applied bias. Note that only macropores on n-type substrates may show a pore spacing significantly exceeding this limit. The regime of stable macropore array formation on n-type Si is indicated by a dot pattern. Type of doping and the formation of current density (in mA/cm^2) are indicated in the legend. From Ref. [5].

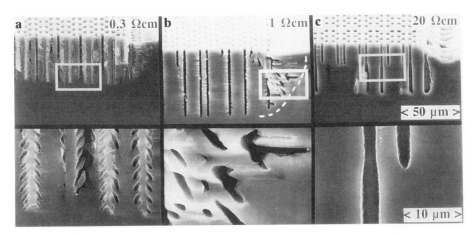

FIGURE 1.7. SEM micrographs of macropore array morphology for the same initiation pattern applied to differently doped n-type electrodes (2.5% HF, 5 mA/cm^2, 2 V). (a) For the highly doped electrode the pitch of the pattern is too coarse, which leads to branching. (c) For the low-doped substrate the pattern is too fine which results in dying of pores. (b) Doping density and pitch are well adjusted in this case and branching is only observed at the border to an unpatterned area (underetching indicated by white dashed line). From Ref. [7].

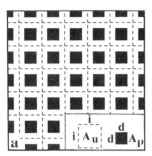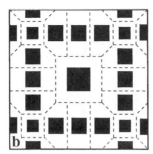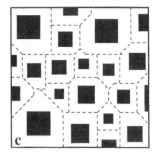

FIGURE 1.8. Sketches showing cross sections of macropore arrays orthogonal to the growth direction (a) for a square, (b) for an ordered and (c) for a random pattern. The pores (black squares) collect holes from the area indicated by the dashed lines. The porosity (the ratio of the black area to the total area) is 0.25 for all patterns.

pattern with a missing pore, a missing row of pores or even two missing rows. Patterns as shown in Figure 1.8b can also be etched. A single pore, however, cannot be etched. It is also possible to enlarge or shrink the pitch of a pattern across the sample surface by a maximum factor of about 3. But a pattern with an abrupt border to an unpatterned area will lead to severe underetching according to Figure 1.7b and random pore formation in the unpatterned area.

A local variation of porosity can be produced by an inhomogeneous illumination intensity. However, any image projected on the backside of the wafer generates a smoothed-out current density distribution on the front side, due to random diffusion of the charge carriers in the bulk. This problem can be reduced if thin wafers or illumination of the front side is used. However, sharp lateral changes of porosity cannot be realized.

Arrays with pore diameters d as small as about 0.3 μm have been realized [7]. The lower limit for the pore diameter of an ordered array is established by breakdown, which leads to light-independent pore growth and spiking. There seems to be no upper limit for the pore diameter, because the formation of 100 μm wide pores has been shown to be feasible [8]. Array porosities may range from 0.01 to close to 1. The porosity, which is controlled by the etching current, determines the ratio between the pore diameter and the pitch of the pore pattern. This means for a square pattern, any pore diameter between one-tenth of the pitch and nearly the pitch can be realized.

The pore diameter can be varied over the length of the pore by a factor up to about 3 for all pores simultaneously by adjusting the current density, as visualized by Figure 1.9. This means the porosity normal to the surface can be varied. The taper of such pore geometries is limited by dying of pores to values below about 30° for a pore diameter decreasing in growth direction, while values in the order of 45° have been realized for an increase of pore diameter in growth direction [9]. Note that narrow bottlenecks will significantly reduce the diffusion in the pore and the formation of deep modulated pores becomes more difficult than the formation of straight pores. Bottlenecks at the pore entrance may result from the transition of the pyramidal etch pit into a pore tip. They can be avoided by an increase of the current density during the first minutes of pore array fabrication.

The pore cross section under stable array formation conditions is usually a rounded square, as shown in Figure 1.10b. Subsequent to electrochemical pore formation, the cross section can be made round by oxidation steps or can be made square by chemical

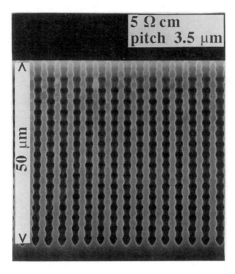

FIGURE 1.9. A sine wave modulation of the etching current versus etching time produces an array of macropores with corresponding modulation of a diameter. From Ref. [7].

etching at RT in aqueous HF or weak alkaline solutions such as diluted KOH or NH$_4$OH. The formation of side pores by branching or spiking, as shown in Figure 1.7a, can be suppressed by an increase of current density or a decrease of doping density, bias or HF concentration. The dying of pores, as shown in Figure 1.7c, is suppressed by an increase of current density, doping density or bias.

The pore length l can be as large as the wafer thickness (up to 1 mm). However, the growth of deep pores requires low electrolyte concentrations, low temperatures and etching times in the order of a day or more, because the etch rate in deep pores is limited by HF diffusion to values in the order of 0.5 μm/min and below. Shorter pores ($l < 0.1$ mm) can be etched much faster (5 μm/min). Under stable etching conditions all pores have the same length. Pore arrays with through-pores can be realized by an increase of the etching current density into the electropolishing regime, which separates a free-standing porous plate from the substrate. Macropores penetrating the whole wafer thickness can be etched as well, however, pore formation becomes unstable in the vicinity of the backside. The formation of dead-end pores and subsequent oxidation and alkaline etchback has found to be technologically favourable.

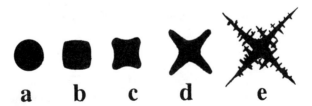

FIGURE 1.10. By an increase of bias or doping density the round (a) or slightly faceted (b) cross section of macropores becomes star shaped by branching (c and d) or spiking (e) along the ⟨100⟩ directions orthogonal to the growth direction.

Another effect which limits the obtainable pore length is characterized by a sudden drop of the growth rate at the pore tip to negligible values and an increase of pore diameter close to the tip. This degradation of pore growth establishes an upper limit for the pore length for a given set of anodization parameters. The fact that pore degradation is delayed for a reduced formation current, which produces conical pores is an indication for a diffusion related phenomenon. The observed dependence of degradation on the concentration of the dissolution product H_2SiF_6 in the electrolyte points to a poisoning of the dissolution reaction. The maximum obtainable pore depth decreases rapidly with increasing HF concentration. This effect has been ascribed to the rate of H_2SiF_6 production being proportional to J_{PS} which again depends exponentially on HF concentration, while the diffusion of H_2SiF_6 is expected to show little dependence on HF concentration [7].

The pore growth direction is along the $\langle 100 \rangle$ direction and toward the source of holes. For the growth of perfect macropores perpendicular to the electrode surface (100)-oriented Si substrates are required. Tilted pore arrays can be etched on substrates with a certain misorientation to the (100) plane. Misorientation, however, enhances the tendency of branching and angles of about $20°$ seem to be an upper limit for unbranched pores.

In conclusion, it can be said that the limits of macropore array formation are in some way complementary to the limitations of plasma etching. The latter technique gives a higher degree of freedom in lateral design, while the freedom in vertical design and the feasible pore aspect ratios is limited.

Today's applications of macropore arrays range from electronic applications such as capacitors to optical filters and biochips.

REFERENCES

[1] T. Martin and K.R. Hebert, Atomic force microscopy study of anodic etching of aluminum, J. Electrochem. Soc. **148**, B101–B109 (2001).
[2] H. Masuda and K. Fukuda, Science **268**, 1466 (1995).
[3] V. Lehmann and H. Föll, Formation mechanism and properties of electrochemically etched trenches in n-type silicon, J. Electrochem. Soc. **137**, 653–659 (1990).
[4] V. Lehmann, R. Stengl and A. Luigart, On the morphology and the electrochemical formation mechanism of mesoporous silicon, Mater. Sci. Eng. B **69–70**, 11–22 (2000).
[5] V. Lehmann, The physics of macropore formation in low doped n-type silicon, J. Electrochem. Soc. **140**, 2836–2843 (1993).
[6] V. Lehmann and S. Rönnebeck, The physics of macropore formation in low doped p-type silicon, J. Electrochem. Soc. **146**, 2968–2975 (1999).
[7] V. Lehmann and U. Grüning, The limits of macropore array fabrication, Thin Sol. Films **297**, 13–17 (1997).
[8] P. Kleinmann, J. Linnros and S. Peterson, Formation of wide and deep pores in silicon by electrochemical etching, Mater. Sci. Eng. B **69–70**, 29–33 (2000).
[9] F. Müller, A. Birner, J. Schilling, U. Gösele, C. Kettner and P. Hänggi, Membranes for micropumps from macroporous silicon, Phys. Status Solidi a **182**, 585 (2000).

2

Macropores in p-Type Silicon

J.-N. Chazalviel and F. Ozanam
Laboratoire de Physique de la Matière Condensée, CNRS-Ecole Polytechnique, Palaiseau, France
E-mail: jean-noel.chazalviel@polytechnique.fr

2.1. INTRODUCTION

Anodization of moderately doped ($N_A \sim 10^{15}$–10^{16} cm^{-3}) p-Si substrates in aqueous or ethanolic HF has long been the most popular method for obtaining good quality microporous silicon [1,2]. The obtained material, exhibiting rather uniform porosity with pore sizes down to the nanometre range, has been the subject of many studies, most of them in the last 10 years being aimed at the understanding of its luminescence properties. Although the formation mechanism of microporous silicon is still a matter of debate, its fabrication can be controlled to a high degree of reproducibility. However, this homogeneous material is actually obtained in a limited doping range of the p-Si substrate, say between 0.1 and a few Ω cm. For highly doped Si (p$^+$), more complex morphologies are obtained, consisting of mesopores growing along the direction parallel to the current lines, with microporous material in between. On the other hand, it had been noted by early workers that a less controlled material is obtained if the resistivity of the starting p-Si is above a few Ω cm. Blackish layers were then observed instead of the coloured films usually obtained with "good" microporous silicon.

More detailed studies have been performed since the mid-1990s. Wehrspohn *et al.* noted that, when porous silicon is prepared from glow-discharge amorphous-hydrogenated silicon (a high-resistivity material), only a very thin layer of microporous material can be formed [3]. When the thickness of the microporous layer reaches a critical value, macropores start growing until they short-circuit the amorphous silicon film. This observation was rationalized in terms of a Laplacian instability: At the interface between two media of different resistivities, the electric current tends to concentrate near the protrusions of the lower resistivity medium. Since the resistivity of hydrogenated amorphous silicon falls in the range 10^4–10^6 Ω cm, the electrolyte is much less resistive than the electrode. This makes the front of the microporous layer unstable on a large scale [3]. Similar growth of macropores was demonstrated by the same authors on highly resistive p-type crystalline silicon, seemingly supporting this simple resistivity argument [4]. However, further investigations by Lehmann and Rönnebeck [5] showed that things are not that simple. Whilst the formation of macropores does cease for substrate resistivities below a critical value, this value does not match the electrolyte resistivity. In parallel, studies in non-aqueous electrolytes were developed by the groups of Kohl [6,7] and Levy-Clément [8–10], and later pursued by the group of Föll [11–14]. Their results show that the interface chemistry plays an important role in orienting the morphology of the porous layer. Here again the resistivity argument appears exceedingly simple, and alternate effects were invoked.

In the following, we will first attempt to summarize the experimental observations and extract the major trends from them. In a second step, we will present the theoretical ideas put forward by the various groups. These ideas will be discussed in the third part. Finally, we will discuss the possibility to form ordered macropore arrays in the light of some recent publications.

2.2. PHENOMENOLOGY

In the following presentation, we will distinguish the macroporous structures formed in aqueous HF and those obtained in non-aqueous solvents. Among the former ones, we will include those obtained in ethanolic medium, since ethanol is just used as a surfactant, and the so-called ethanolic media are still mostly aqueous. However, we should also keep in mind that the electrolytes made from non-aqueous solvents often include significant amounts of water incorporated with the hydrofluoric acid, so that the boundary between aqueous and non-aqueous media is not really clear cut. We will nevertheless stick to it for the ease of presentation.

2.2.1. Macropore Formation in Aqueous (and Ethanolic) HF Electrolytes

Macropores are obtained in aqueous electrolytes for p-Si resistivities above a critical value, on the order of a few Ω cm. This occurs essentially in the same range of current densities as that leading to microporous-silicon formation. These conditions are recalled in Figure 2.1. Porous silicon is formed in the rising part of the current–potential curve. When current density reaches a critical value J_c, the surface becomes covered with an oxide film [15], and electropolishing takes place. The value of J_c increases with increasing concentration of HF in the electrolyte [16–21], but it also depends on solution stirring [17,19–21], temperature [18,21] and crystallographic orientation of the silicon substrate [16,18]. When anodization is carried out at a fixed current density $J < J_c$ without a prepatterning of the Si surface, microporous silicon is formed first. Macropores appear when the thickness of the microporous layer exceeds a few micrometres. This preliminary stage is sometimes referred to as the nucleation stage (though micropores and mesopores are already present at a much earlier stage). For dilute HF electrolytes or prolonged anodization times, this microporous nucleation layer is sometimes not observed on the

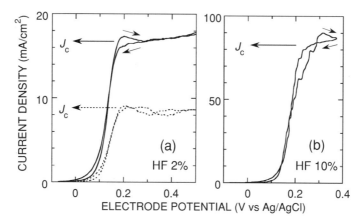

FIGURE 2.1. Voltammograms of p-Si (111) in typical aqueous HF electrolytes: (a) 2% HF (HF:H$_2$O:ethanol 2:68:30 (solid line) and 2:98:0 (dotted line)) and (b) 10% HF (HF:H$_2$O:ethanol 10:60:30). The potential scale has been corrected for ohmic drop in the silicon and the electrolyte. Recorded on a rotating disc electrode. Rotation rate 300 rpm. Temperature 18 °C. The fluctuations are due to the formation of bubbles at the electrode surface. The electropolishing current J_c is seen to increase with increasing HF concentration, but it also depends on ethanol concentration, temperature, and stirring of the electrolyte (or electrode rotation rate).

obtained samples. This is clearly due to its chemical dissolution while the macroporous layer thickens. Using short anodization times shows that it is always present at the beginning of the anodization.

2.2.1.1. Current-Line-Driven Pores and Crystallography-Driven Pores. Observation of the morphologies obtained reveals that there are two rather distinct classes of macropores, as can be seen from Figure 2.2.

- Macropores of the first class exhibit rounded bottoms and somewhat meandering walls. Their average orientation is normal to the sample surface, irrespective of the crystal orientation, and they exhibit smooth bending near the edge of the anodized area. These macropores actually appear to be filled with microporous silicon.
- On the opposite, macropores of the second class tend to form (111) facets at their bottoms and to have (110)-oriented walls. In cross section, they exhibit a more or less rounded square shape when grown along (100), and a more or less rounded triangular shape when grown along (111). No silicon is left in these pores when they form, except possibly for some microporous silicon coating their walls. For these pores, the (100) direction appears as a preferred growth direction: growth occurs along (100) even if the surface is oriented at some angle from that direction. However, the preference for (100) appears less marked than in the case of macroporous n-Si. Also, if the substrate is not free from mechanical damage, these pores may grow preferentially along dislocations lines.

These two kinds of pores are sometimes referred to as "current-line-driven pores" and "crystallography-driven pores", respectively, and we will adopt these terms in the following. For a given electrolyte composition and a given silicon doping, a transition

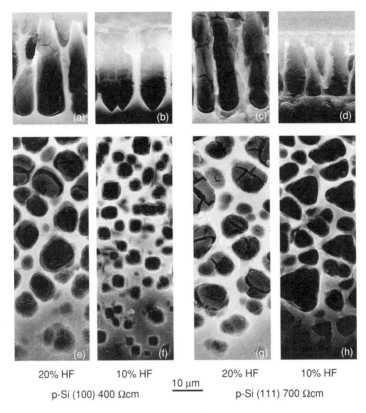

FIGURE 2.2. Crystallography-driven pores versus current-line-driven pores. Side views (a–d) and top views on a bevelled edge near the pore bottoms (e–h) of macropores formed on p-Si at a current density of 30 mA/cm^2 (anodization time 20 minutes), in various experimental conditions, indicated at the bottom of the figure. Note the porous-silicon filling and the rounded shape of current-line-driven pores (a, e, c, g), and the change in shape of crystallography-driven pores depending on the orientation of the surface (squared shape on (100) (b, f) and triangular shape on (111) (d, h)).

from current-line-driven pores to crystallography-driven pores is observed as current density is increased. There seems to be more or less agreement among the groups that crystallography-driven pores are obtained when the pore bottoms are under electropolishing conditions. The known variation of the electropolishing current density as a function of surface orientation [16,18] would then explain the anisotropy of pore growth. However, though this interpretation is plausible, there is as yet no quantitative proof of the relationship between the two phenomena, nor is it clear whether the moderate anisotropy of the electropolishing current is sufficient to account for the sharp angles observed in the pore shape.

When current density is further increased, pore formation stops, and a wavy surface is formed instead. Surprisingly, Lehmann and Rönnebeck have reported that this change of regime occurs for a current significantly smaller than the nominal value of the electropolishing plateau [5]. However, this limit is accessible only in electrolytes with a rather low HF concentration, where chemical dissolution of silicon is non-negligible,

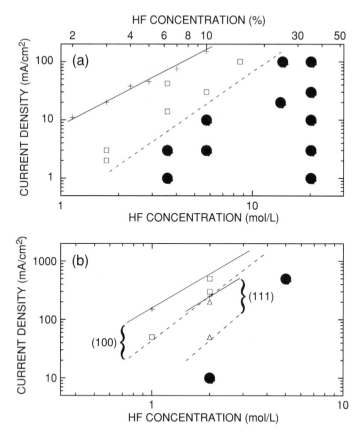

FIGURE 2.3. Morphological map of macropores on p-Si: (a) in aqueous (ethanolic) HF electrolyte and (b) in acetonitrile electrolyte (with 2.4 M H_2O). The crosses and the solid lines represent the electropolishing regime. The regime of current-line-driven pores (black circles) and that of crystallography-driven pores (squares for (100), triangles for (111)) are separated by a dashed line. The data in (a) are for (100) orientation. Note the difference between (100) and (111) in (b). Data are taken from [5,8,9,22,23].

and HF depletion at the pore bottoms may be important. We then regard as plausible that in this regime a highly porous structure is actually formed, but its slow chemical dissolution or mechanical breaking may be responsible for the mismatch in the current densities. Figure 2.3a summarizes results collected from various sources [5,22,23], giving a map of the two morphological regimes in the HF concentration/current-density plane.

2.2.1.2. Characteristic Sizes. The characteristic sizes of the macroporous structure, that is, the average pore diameter and wall thickness, both increase with increasing silicon resistivity, spanning a typical range from 1 μm to 10 μm, as the resistivity is increased from 10 Ω cm to 1000 Ω cm and above. The two quantities vary about proportionally to $\rho^{1/2}$, where ρ is silicon resistivity. As said above, for resistivities below a few Ω cm, there is no longer formation of a macrostructure, but rather a uniform micro(meso)porous layer grows steadily, up to thicknesses as large as a few hundred micrometres.

The growth rate of the macropores is about the same as that of microporous silicon. Especially, it is about proportional to the applied current density [5]. This feature stands in contrast to macropore growth on n-Si, which occurs at a constant rate, set by the value of the electropolishing current density. The fact that macropores on p-Si can grow much more slowly than on n-Si clearly indicates that current density at the pore bottoms may be much smaller for p-Si than for n-Si. This is especially true in the regime of current-line-driven pores.

2.2.2. Macropore Formation in Non-Aqueous Electrolytes

It is a common rule that the electrochemical behaviour of an electrode in a non-aqueous electrolyte is often governed by the small amount of water present in that electrolyte. Formation of porous silicon in non-aqueous electrolytes seems indeed to follow that rule. It seems that the only investigations on the behaviour of silicon in anhydrous HF electrolytes were made by the group of Kohl [6,7], using acetonitrile (ACN) as the solvent. For p-type silicon, the results were that no microporous silicon is ever formed. For (100) orientation, crystallography-driven macropores were obtained. On (111), only pyramidal etch pits were observed. However, if some water is present in the organic electrolyte, behaviours reminiscent of those observed in aqueous HF appear. Many of such experiments have been made in various solvents, the most used ones being ACN, propylene carbonate (PC), dimethylsulfoxide (DMSO) and dimethylformamide (DMF).

Among those solvents, ACN is probably the best documented case. In anhydrous ACN + HF solutions, the voltammogram of p-Si exhibits a continuous rise [6], without any limitation such as the electropolishing plateau of Figure 2.1. However, in the presence of water, such a limitation appears, due to the formation of an oxide when the current density reaches a critical value [6,8]. In such electrolytes, a microporous nucleation layer is formed, just as in aqueous electrolytes. When the current density is increased, the same morphology sequence as in aqueous electrolytes is observed: current-line-driven macropores (filled with microporous silicon), then crystallography-driven macropores and finally electropolishing. A map of the morphological regimes may be drawn from the few data available in the literature, and is shown as Figure 2.3b, here for a water concentration of 2.4 M [8]. Note the similarity with Figure 2.3a, except that the boundaries between the different regimes occur at lower HF concentration and/or higher current densities than in the case of aqueous electrolytes. The average diameter of the macropores and the average wall thickness have been found to exhibit a slight decrease upon increasing current density [8] (though the observed increase in pore size with increasing porous-layer thickness makes it difficult to extract a single figure for the average pore diameter). Adding increasing amounts of water to the electrolyte results in an increase of pore diameter, and may ultimately result in electropolishing, because the electropolishing current density J_c decreases as a function of water concentration in the H_2O/ACN mixtures (a similar variation of J_c is observed for H_2O/ethanol mixtures, as can be seen from Figure 2.1a). Finally, it was found that, as in the case of aqueous electrolytes, macropores can be obtained only for p-Si resistivities above a few Ω cm [8].

When other solvents are used, the characteristic sizes of the porous structure appear to depend on the solvent to a rather weak extent (see Table 2.1 of [10]). However, the morphological maps exhibit significant variations. Whilst using PC leads to results closely similar to those obtained with ACN, a wider range for crystallography-driven pores is observed when using DMF and DMSO [9,10,13,14]. For these solvents, macropores can be formed down to very small current densities, and the crystallographic effects appear strongly enhanced: not only do the pore bottoms exhibit (111) facets, but the preference for the (100) growth direction is so strong that (100)-oriented pores are obtained even when starting from a (111) Si surface (see Figure 2.4) [13,14]. To a lesser extent, the (113) directions also appear as preferred growth directions (see Figure 2.4) [13,14]. These features, reminiscent of those observed for n-Si, are observed for p-Si

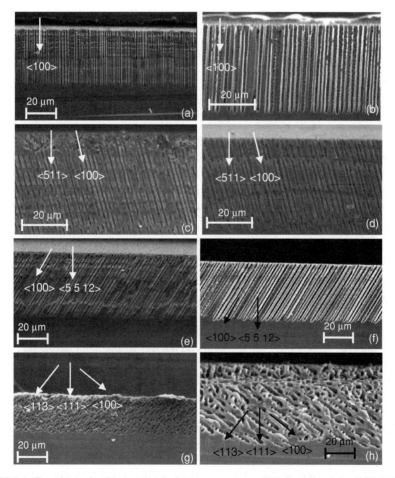

FIGURE 2.4. Crystallography-driven pores obtained in non-aqueous HF electrolytes (after [14]). DMSO + 4% H_2O + 4% HF (a, c, e, g) and DMF + 4% H_2O + 4% HF (b, d, f, h). p-Si resistivity 10–20 Ω cm. Current density 2 mA/cm^2. Note the marked preference for ⟨100⟩ and ⟨113⟩ growth directions. Reproduced by permission of the Electrochemical Society.

only in these peculiar non-aqueous solvents. Finally, in DMF and DMSO, macropores have been formed down to silicon resistivities of 0.2 Ω cm [10].

2.3. THEORY

In this section, we will try to present the various ideas that have been put forward in order to rationalize the above observations. The presentation will be organized with respect to the classes of ideas, which closely map the various different groups. The discussion section will be organized with respect to the different phenomena and systems.

2.3.1. Role of the Space Charge

Following ideas first proposed by Lehmann and Föll for macropore formation on n-Si [24], Lehmann and Rönnebeck have considered the plausible role of the space charge in the formation of macropores on p-Si [5]. The idea is that silicon dissolution is governed by the flow of holes reaching the surface. Since the surface is under weak depletion conditions, the holes have to overcome a Schottky barrier. According to Lehmann and Rönnebeck [5], the overall rate of hole transfer is limited by the diffusion velocity over the barrier. This quantity is proportional to the electric field inside the barrier, that is, inversely proportional to the barrier thickness. At the pore tips, due to interface curvature, the space-charge layer is thinner, which makes the hole diffusion velocity higher at these locations. This results in an increased dissolution rate at the pore tips. In line with these ideas, the wall thickness is expected to be determined by the thickness of the depletion layer: the maximum thickness so that a wall is fully depleted is just $2w_{SC}$, where w_{SC} is the usual depletion-layer width. However, to this point the present theory does not give a prediction on characteristic pore diameter.

2.3.2. Chemical Effects

The proposal of Lehmann and Rönnebeck allows one to understand the variation of the wall thickness with silicon doping. However, from the observation that very different morphologies may be obtained when changing the solvent, it is clear that chemical effects come into play. For example, as pointed out by Kohl et al. [6,7], in anhydrous ACN electrolyte, there cannot be any oxide on the Si surface, and dissolution proceeds by the formation of SiF bonds only. This argument obviously does not hold when some water is present. However, many other chemical factors may play a role.

Levy-Clément et al. have underlined the fact that the dissolution rate may not be limited by hole supply to the interface, in contrast to Lehmann and Rönnebeck's assumption, but also by interface kinetics. When one writes the chemical steps of the Si dissolution, it is clear that the intermediate species may interact with the solvent, and the rates of each step may depend on the polar character of the solvent, its ability to solvate nucleophilic anions and the solubility of the dissolution products [10]. Furthermore, the band bending, which plays a key role in Lehmann and Rönnebeck's approach, may be solvent dependent: for different solvents, the applied potential may divide in different ways between the semiconductor space charge and the Helmholtz layer. An interesting point is that DMF and DMSO, which allow to grow spectacular crystallography-driven pores, belong to the same class of aprotic protophilic solvents [10]. However, the relationship between these two experimental facts is not yet clear.

On the other hand, the group of Föll has addressed the question of anisotropic pore growth in more detail, and has proposed a new model, which among other results may account for these effects. The model rests on the hypothesis that, on a local scale, dissolution takes place through cyclic "current bursts" [25]. A cycle begins with a current increase leading to local Si dissolution, then the reactants get exhausted and the surface gets oxidized. When the oxide gets thick enough, the current vanishes, the oxide dissolves, and the surface gets rehydrogenated and passivated, till a new cycle starts [25]. The hydrogen repassivation step would be that responsible for the anisotropic effects. At the core of the model is the idea of a competition between oxidation and hydrogenation, a competition that would lead to unstable behaviour on a local scale. The authors suggest that macropores can appear when there is sufficient oxidation to balance direct dissolution. Since oxidation is considered to be an isotropic process, more anisotropic pores are expected in "less oxidizing" solvents, or in circumstances where there is less oxygen and more hydrogen. As a support to their model, the authors report that adding diethyleneglycol (supplying hydrogen) to ACN makes the pores more regular [11].

2.3.3. Linear Stability Analysis (LSA) Approach

A problem with many of the above theories is that they are more qualitative than quantitative. Hence, they are rather difficult to either prove or disprove. In an attempt to make quantitative predictions, several groups have tried to use linear stability analysis (LSA) in order to assess their ideas. Linear stability analysis is a general method to study the behaviour of the interface between two phases [26]. It is used especially in the study of growth phenomena. It consists in studying the stability of a flat interface. For that purpose, a small sinewave perturbation of the flat interface is assumed, of amplitude δ and wave vector q (see Figure 2.5), and the equations of the proposed model are used to explore the time evolution of that perturbation. Since the perturbation is taken infinitely small, the equations can be linearized and the evolution problem can often be solved with a limited amount of mathematics. The evolution of δ comes out generally of the form $d\delta/dt = \beta(q) \times \delta$. If $\beta(q) < 0$, the perturbation tends to damp out exponentially: the interface is stable with respect to a perturbation of wave vector q. If $\beta(q) > 0$,

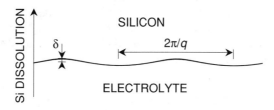

FIGURE 2.5. Principle of linear stability analysis. An initially planar interface is perturbed by a small sinewave perturbation of amplitude δ and wave vector q, and the evolution of δ as a function of time is studied.

the perturbation increases exponentially with time, until the linear approximation holds no longer. The wave vector q_{max} corresponding to the maximum value of β gives the characteristic scale of the perturbation that will appear first. Therefore, this method is a valuable tool to predict the onset of instability of an interface. However, in principle, it cannot predict its behaviour beyond the linear regime.

Linear stability analysis has been used by Kang and Jorné for the study of macroporous-silicon formation on n-Si [27], and later by Valance for the study of porous-silicon formation on n-Si and p-Si [28,29]. However, the latter results were rather at variance with the experimental data. Recently, our group has reconsidered the case of p-Si, and fair agreement was obtained [30,31,23]. Our model will be described here in some detail. It rests on the classical assumption that silicon dissolution is governed by the reaction of the first hole at the interface. Our ingredients are essentially an extension of the space-charge model, aimed at incorporating other effects, and include the following:

1. Near the maxima of the sinewave (depressions of the Si surface), the thinning of the space-charge layer results in an increase of the hole diffusion velocity $v_D = \mu E_S$, where μ is hole mobility and E_S interface electric field. This is nothing but the LSA version of Lehmann and Rönnebeck's argument [5,30].

2. At the same locations, the increased interface electric field E_S results in an increased Helmholtz potential drop [32], and a decrease in band bending (Schottky-barrier lowering), given by $\Delta\Phi_{SC} = -e\varepsilon\varepsilon_0 E_S/C_H$, where e is elementary charge, C_H is Helmholtz capacitance and $\varepsilon\varepsilon_0$ is the permittivity of silicon. This effect was not taken into account by previous investigators, especially Lehmann and Rönnebeck.

3. The reactivity of the interface is characterized by a reaction velocity v_R, which occurs as a process in series with hole transport through the space-charge layer. Namely, the current is taken as $J = -eN_A v \exp(-\Phi_{SC}/k_B T)$, where $v^{-1} = v_D^{-1} + v_R^{-1}$, N_A is the acceptor concentration, Φ_{SC} is the band bending at the applied potential considered, T is the absolute temperature and k_B is the Boltzmann's constant. The reaction velocity v_R is an important feature, as it bears all the chemical information in the model. It is assumed to vary with surface curvature as $v_R = v_R^0 (1 + \kappa a^2)$, where κ is curvature (taken positive for a surface protrusion) and a a characteristic interatomic length. This form provides a fair description of the higher surface reactivity at the more open positions available on a silicon protrusion, and the lower reactivity due to steric hindrance at the depressions. However, by construction it assumes direct hole transfer at the interface. Especially, it takes into account neither the possible presence of surface states nor that of an oxide layer. Therefore, it cannot apply to the electropolishing regime and cannot account for crystallographic effects.

4. On a large scale (small q), the potential distribution is affected by the resistivities of the silicon and the electrolyte.

Working out this model leads to a $\beta(q)$ function always positive in a wide range of q's. Figure 2.6 shows a typical plot of $\alpha(q) = \beta(q)/V$, where V is the average velocity of the interface (knowledge of α is equivalent to that of β, but α is more convenient here as it is homogeneous to an inverse length).

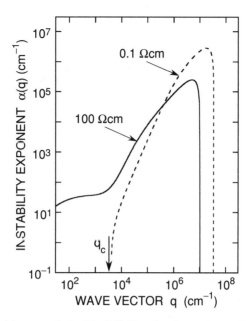

FIGURE 2.6. Typical $\alpha(q)$ curves obtained in [31]. Note that the maximum of α occurs for $1/q$ in the nanometre range, but the behaviour at small q is strongly dependent on silicon resistivity.

Near its maximum, α is approximately given by $Sq - q^2 a v_D/(v_R + v_D)$, where $S = v_R/(v_R + v_D) + e\varepsilon\varepsilon_0 E_S/k_B T C_H$ is a dimensionless parameter, on the order of 0.1–1, which arises from the destabilizing effects 1 and 2 [30]. The scale where the first instability of the interface occurs is then on the order of $1/q_{max} \sim a/S$, in the nanometre range. The model then accounts for the initial stages of the formation of microporous silicon, giving a correct order of magnitude for the scale of the structures in that material [31].

In principle, LSA does not allow one to predict what occurs beyond the linear regime. Here, however, information on the larger scales can still be obtained: when a microporous layer is formed, due to depletion of the semiconductor in the microporous structure, the physical properties of the layer are essentially those of the electrolyte inside the pores. The above LSA approach can then be used as it stands, for the study of the stability of the front of the porous layer. The existence of a wide range of q's below q_{max}, where α is positive, shows that the front will develop instabilities at increasingly large scales (q^{-1} increasing with α^{-1}, which is itself on the order of the porous-layer thickness Δ), thereby forming mesopores and macropores filled with microporous silicon.

Interestingly, depending on the relative resistivities of the semiconductor ρ_s and the electrolyte ρ_e, the range of positive α's extends down to zero q (for $\rho_s > \rho_e$) or stops at a critical value q_c (for $\rho_s < \rho_e$) (see Figure 2.6). From this result, it was earlier concluded that macropore formation is governed by the resistivity ratio ρ_e/ρ_s [3,4]. However, this conclusion was disproved by experiment [5,8–10]. As a matter of fact, an inspection of Figure 2.6 shows that the value of q_c is so low and corresponds to values of α so small

SILICON

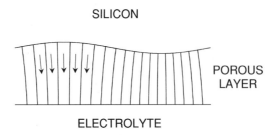

POROUS
LAYER

ELECTROLYTE

FIGURE 2.7. Stabilization of the pore front by the resistivity of the electrolyte. The pores grow in the direction perpendicular to the front. Since the pore walls are insulating, the current feeding a protruding part of the front has to flow through a narrowed region of the electrolyte, which strongly enhances the ohmic drop associated with that region.

that this regime is hardly reached in practice. In practical conditions, α values down to 10^2–10^4 cm^{-1} may be operative (depending on the porous-layer thickness $\Delta \sim \alpha^{-1}$). In this regime, the model does predict the formation of macropores of increasing size, with characteristics weakly dependent on the ρ_e/ρ_s ratio. Therefore, the actual question is not about the origin of macropores, but rather about their disappearance at low resistivities, and also about the observed morphology: why do the macropores grow parallel, rather than exhibiting a hierarchy of increasingly large pores, as would be expected from the positive value of α up to large scales? These two questions were addressed in a recent work [31,23], which is summarized hereafter.

Treating the microporous layer as an effective medium of isotropic resistivity ρ_e is not appropriate: since the walls are insulating, the preferential orientation of the pores parallel to the growth direction makes the resistivity anisotropic. Furthermore, the pores tend to grow perpendicular to the local front surface. Hence, around a protrusion of the porous layer, the increased interface area is fed by a narrower bunch of pores (see Figure 2.7). This makes the electrolyte resistivity play a much more important role than would be the case for an isotropic medium. A detailed calculation shows that a stabilizing contribution to α arises from that effect, and this contribution increases with increasing layer thickness. At early growth times, a hierarchical array of macropores is then expected, till this stabilizing contribution counterbalances the destabilizing contribution of Figure 2.6. The pores existing at this stage then continue their growth as a stable parallel array. This change of regime is predicted to occur for a characteristic size $1/q_c^*$ [23]:

$$\frac{1}{q_c^*} \approx \left[\lambda^3 S^3 \frac{k_B T}{e J \rho_e} \right]^{1/4} \tag{2.1}$$

where λ is the ratio Φ_{SC}/eE_S (i.e., $\lambda \sim w_{SC}$). The characteristic length π/q_c^* is the characteristic pore diameter above which parallel pore growth is predicted to occur. This prediction should apply to the regime of current-line-driven macropores.

The last intriguing point is the observation that no macropores are formed below a certain resistivity of the silicon sample. A proposal can be made from the fact that, when doping increases, the average space-charge thickness decreases as $N_A^{-1/2}$, whereas the average interimpurity distance decreases as $N_A^{-1/3}$. The two quantities cross smoothly for $N_A \sim 10^{18}$ cm^{-3}, but they are pretty much of the same order of magnitude already for

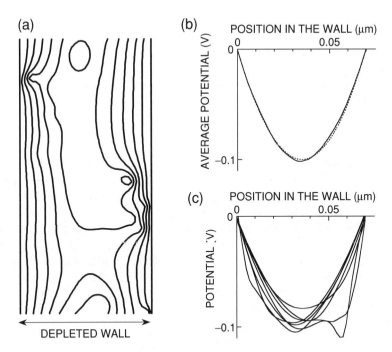

FIGURE 2.8. Fluctuation effects due to the small number of impurities in a depleted wall when doping increases. Here, simulation of a portion of depleted wall, of thickness $2w_{SC} = 0.07$ μm and surface area $4w_{SC} \times 4w_{SC}$, for a doping of 10^{17} cm^{-3} (there are 140 ionized impurities in this volume). (a) two-dimensional map in a cross section of the wall. (b) Potential averaged in the directions parallel to the wall. (c) Potential along a few straight lines perpendicular to the wall. Although the potential averaged over the directions parallel to the wall is fairly close to the classical parabolic profile (dotted line in (b)), the map of the potential in a cross section of the wall ((a) or (c)) exhibits strong deviations from this profile. This may render the wall permeable to holes, so that the assumption of an insulating wall breaks down (after [23]).

$N_A \gtrsim 10^{16}$ cm^{-3}. As a result, the number of ionized impurities when crossing a depleted wall is small, and fluctuation effects may become important, leading to conduction paths through the wall (see Figure 2.8). We have made a numerical simulation of a depleted wall with a band bending of 0.1 eV, which is the typical estimated band bending in the regime of porous-silicon formation in aqueous (ethanolic) HF. The insulating character of the wall was determined from the energy of the lowest acceptor state in a parallelepipedic portion of the wall. As fluctuation effects become important, the wall remains insulating for thicknesses smaller and smaller as compared to the naive $2w_{SC} \propto N_A^{-1/2}$ law. The conclusion is that a depleted wall can no longer remain insulating for a critical doping in excess of a few 10^{16} cm^{-3} [23]. A major consequence of the loss of this insulating character is that the anisotropy of the resistivity of the porous layer disappears, the above reasoning breaks down, and no parallel macropore growth is to be expected. As first suggested by Lehmann and Rönnebeck (though for reasons that are in our opinion incorrect), higher dopings result in the growth of a mesoporous layer. Interestingly, this critical doping is expected to strongly increase for higher band bendings [23].

2.4. DISCUSSION

Here, we will put the emphasis on the possibilities to get predictive quantitative information on characteristic pore sizes and morphologies. This will be less of a review and more of the authors' personal views. The discussion will be organized according to the two classes of pores.

2.4.1. Crystallography-Driven Pores

If we accept that crystallography-driven pores are under electropolishing conditions at their bottoms, Lehmann and Rönnebeck's ideas can be turned into quantitative predictions on characteristic pore sizes. As stated in Section 2.3.1, the average wall thickness is expected to be given by $2w_{SC}$ [5]. Now, if the current density at the pore bottoms is set to the electropolishing value J_c, one can write $P = J/J_c$, where P is porosity. This argument is similar to that given for the case of n-Si [24,18]. For a given pore geometry, there is a relationship between porosity, pore diameter and wall thickness. This relationship is not too much dependent on pore geometry. Taking either a honeycomb lattice of hexagonal pores or a square lattice of square pores, porosity is $P = [R/(R + w_{SC})]^2$, where R is pore "radius" (here apothem of the polygon). Hence, we get the pore diameter d (a slightly modified form of original Lehmann's formula) [18]:

$$d = \frac{2w_{SC}}{P^{-1/2} - 1} = \frac{2w_{SC}}{(J_c/J)^{1/2} - 1}.$$

(2.2)

The observed increase in wall thickness as a function of doping is in fair agreement with Lehmann and Rönnebeck's prediction. Lust and Levy-Clément have pointed out that there does not seem to be a minimum wall thickness [10]. However, this is not really surprising in view of the statistical dopant fluctuations discussed in Section 2.3.3 and the disordered macropore structure. As a matter of fact, the average (rather than minimum) wall thickness seems to match the $2w_{SC}$ prediction reasonably well. Interestingly, this prediction appears better verified on p-Si than on n-Si (where minority-carrier diffusion may lead to pore spacings well in excess of $2w_{SC}$) [5,33]. Also, the increase in pore diameter before approaching the transition to electropolishing is well accounted for by Equation (2.2). Finally, we regard as very plausible that the anisotropy of the electropolishing current bears the explanation for the anisotropy of pore shape and pore growth direction. In the detail, the difference between "anisotropy of the electropolishing current" and the model of Föll *et al.* (anisotropy of the kinetics of rehydrogenation) may be mostly a matter of vocabulary. However, anisotropy of the electropolishing current may also result from factors not taken into account in Föll's model (e.g., role of SiOH species), and the oscillatory nature of the current on the microscopic scale does not appear as a mandatory requirement for electropolishing to take place. Nevertheless, a careful study of the electropolishing current as a function of crystallographic orientation would be necessary in order to quantitatively assess the role of its anisotropy in the growth of macropores. In the detail, one may also wonder which value should be taken for J_c in Equation (2.2). For example, if pores are growing along (100) with (111) facetted bottoms, J_c should plausibly be taken as $\sqrt{3} \times J_c(111)$ rather than as $J_c(100)$, but it is not clear either whether $J_c(100)$ can actually be determined by a direct measurement.

These problems may be related to the discrepancy observed by Lehmann and Rönnebeck between the electropolishing current and that corresponding to disappearance of the macropores.

2.4.2. Current-Line-Driven Pores

For the case of current-line-driven pores, the space-charge argument is still operative for predicting the average wall thickness, a prediction which remains in fair agreement with observations (see Figure 2.9). However, the pore diameter can no longer be determined by the electropolishing condition, and the only available theoretical prediction is that stemming from the LSA study of the pore front (Equation (2.1)). Let us recall that the basic mechanism at work in this theory is the current limitation due to the series resistance of the electrolyte along the pores. Though the early proposed criterion on the semiconductor to electrolyte resistivity ratio has been disproved by experiment, the theory looks attractive, as it accounts for the observations on the nucleation stage and the progressive increase in pore size before reaching a steady state of parallel growth.

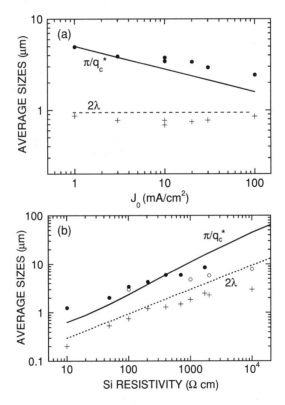

FIGURE 2.9. Comparison of characteristic macropore sizes in the current-line-driven regime, when changing current density (a) and silicon doping (b), in 35% ethanolic HF (HF:H_2O:ethanol 35:35:30 by volume), after [23]. The hollow symbols refer to data obtained in 25% ethanolic HF. The solid line is from Equation (2.1) and the dotted line is $2w_{SC}$. In (a), $\rho_e = 100\,\Omega$ cm. In (b), $J = 10$ mA/cm^2.

Furthermore, electrolyte resistivity must still play a role at large scale, especially in view of the wall depletion, which constrains the current lines inside the pores.

Figure 2.9 shows a comparison of experiment with the theoretical predictions (wall thickness given by $2w_{SC}$, and pore diameter given by Equation (2.1)), when varying current density and silicon resistivity, for a 35% ethanolic HF electrolyte. The agreement is seen to be fair. The deviations between the theory and experiment for the highest resistivities may be due to failure of the assumption that w_{SC} scales as $\rho_s^{1/2}$, since highly resistive samples are generally compensated. Figure 2.10 shows similar comparisons between theory and experiment when electrolyte composition is changed. Such changes may result in changing the reaction velocity v_R (assumed proportional to HF concentration) and the electrolyte resistivity ρ_e (determined experimentally). Here, again the agreement is fine. Note that, for Figure 2.10a, S was adjusted due to the lack of information on Φ_{SC} and C_H in ACN (the resulting value $S = 0.85$ appearing somewhat large), but Figure 2.10b was obtained without any adjustable parameter.

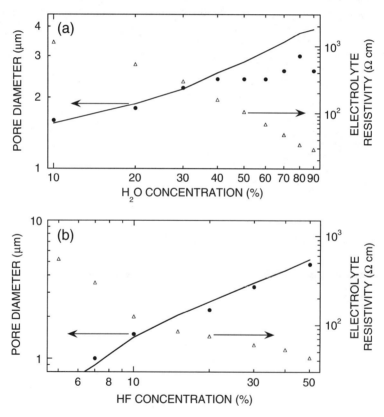

FIGURE 2.10. Comparison of characteristic macropore sizes in the current-line-driven pore regime, when changing the electrolyte. Experiment (black circles) and theory (curves) using measured electrolyte resistivity (hollow triangles). (a) Changing H_2O concentration in acetonitrile/2 M HF electrolyte. $\rho_s = 10–15\,\Omega$ cm, $J = 10$ mA/cm^2. Data from [8]. Theory from Equation (2.1) with $\lambda = 0.165\,\mu$m, $S = 0.85$. (b) Changing HF concentration in mixtures HF:H_2O:ethyleneglycol x:x:$100 - 2x$.$\rho_s = 1500\,\Omega$ cm, $J = 10$ mA/cm^2. Data from [23]. Theory from Equation (2.1) with $\lambda = 1\,\mu$m, $\Phi_{SC} = 0.1$ eV (whence $v_D = 4.7\,10^5$ cm/s) and v_R (cm/s) $= 3\,10^3 \times$ [HF%].

Finally, the observed disappearance of macropores when the resistivity of p-Si becomes lower than a few Ω cm is fairly well accounted for by the loss of the insulating character of the walls due to statistical fluctuations of the dopant concentration (see Figure 2.8). In conclusion, the theory of [23] could certainly be refined by taking into account, e.g., the possible effect of surface states, HF depletion at the pore bottoms, and quantum confinement effects [34]. However, the above facts give the feeling that the formation of current-line-driven macropores in p-Si is fairly well accounted for by that theory in its present form.

2.4.3. Morphological Map

If Equation (2.1) accounts for the average diameter of current-line-driven macropores and Equation (2.2) for that of crystallography-driven macropores, the change from current-line-driven pores to crystallography-driven pores is expected to occur when the second member of Equation (2.1) equals that of Equation (2.2). Physically, this means that the two stabilizing effects (series resistance of the macropores and electropolishing at the pore bottoms) are of comparable strength. If the two quantities in Equations (2.1) and (2.2) are plotted as a function of J (see Figure 2.11), it is seen that the two curves cross somewhere below the electropolishing current density J_c (which is an increasing function of HF concentration). At current densities lower than the crossover value, Equation (2.1) predicts a larger pore size than Equation (2.2): therefore, electropolishing at the pore bottoms is not reached, and current-line-driven pores are predicted, with a diameter of the form $A J^{-1/4}$, where A is a constant which can be extracted from Equation (2.1). At current densities higher than the crossover value, the pore size predicted by Equation (2.1) would lead to current densities at the pore bottoms higher than the electropolishing current. Therefore, the limitation by electropolishing takes over that by the series resistance of the electrolyte, and crystallography-driven pores are expected. So, one may generally expect that crystallography-driven pores will be

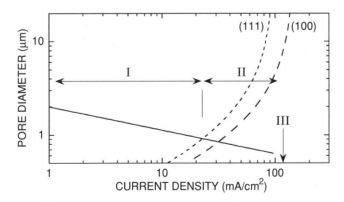

FIGURE 2.11. Predicted crossing between current-line-driven pores (Equation (2.1), solid line) and crystallography-driven pores (Equation (2.2), dotted line for (111), assuming $J_c(111) = 100$ mA/cm^2, and dashed line for (100), assuming $J_c(100) = 150$ mA/cm^2). When current is increased, one successively encounters current-line-driven pores (I), crystallography-driven pores (II), and electropolishing (III). Here, the curves have been calculated for 10% ethanolic HF (HF:H$_2$O:ethanol 10:50:30, $\rho_e = 20\,\Omega$ cm), and $\rho_s = 100\,\Omega$ cm. Note the fair agreement with Figure 2.3a.

observed in a limited range of current densities, just below the onset of electropolishing. In practice, this range is probably widened by the change in dissolution valence from ~ 2 to ~ 4 when entering the electropolishing regime [5,35,36]. Also, it will be wider if the anisotropy of the electropolishing current is larger, and/or if the prefactor A from Equation (2.1) is smaller. This may be the case if ρ_e is large and/or if S is small, which might be favoured by a small value of the reaction velocity υ_R.

At first sight, the morphologies observed for non-aqueous solvents might give the feeling that things are quite different from the case of aqueous HF [14]. However, this feeling is largely due to the fact that most data in the literature are obtained at a given current density. If the full morphological map is considered (i.e., in the HF concentration/current-density plane), there does not seem to be real qualitative differences between aqueous and non-aqueous media. In ACN, it has been observed that the boundaries between the different morphological regimes occur at a lower HF concentration and/or higher current density than in aqueous electrolytes (see Figure 2.3). This effect is clearly due to the lower water concentration and weaker probability of forming an oxide (whence a higher electropolishing current density). On the opposite, the lower electropolishing current density observed in DMF and DMSO, plausibly due to the lower solubility of fluorosilicates in these solvents, leads to a shifting of the morphological map towards the low current densities. The wider range of crystallography-driven pores observed in these solvents may be due to a stronger anisotropy of the electropolishing current, and/or to a smaller value of the prefactor in Equation (2.1).

Finally, the lower resistivity limit down to which macropores can be grown has been found to be smaller in DMF and DMSO than in water, ACN or PC. According to the theory of Section 2.3.3, this might be due to a larger value of the band bending Φ_{SC} under porous-silicon formation conditions in these solvents. Unfortunately, there is as yet no published determination of Φ_{SC} in these systems. However, for anodization in an aqueous electrolyte, it has been reported that addition of a cationic surfactant improves the "quality" of the pores dramatically, whereas addition of an anionic or a neutral surfactant has no effect or even detrimental effects [37]. Adsorption of cationic species is expected to shift the flatband potential positively, that is, to increase the band bending at the p-Si/electrolyte interface. One may then infer that the beneficial effect of a cationic surfactant is indeed related to such an increase of the band bending, which further supports the theory of [23].

2.5. ORDERED MACROPORE ARRAYS

Formation of ordered macropore arrays has long been a tantalizing goal, in view of the possible applications to the fabrication of microstructures for electronic, microme-chanical or optical applications. Crystallography-driven pores appear as the best choice for realizing regular structures, since their growth is controlled by the crystallographic directions and may therefore be perfectly rectilinear. However, a perfectly regular array will not tend to form spontaneously: the pores exhibit short-range order but no sponta-neous long-range ordering. Rather, the mechanism of stabilization by the resistivity of the electrolyte tends to make the pores grow with the same cross section as they started. For obtaining regular arrays, prepatterning is then necessary.

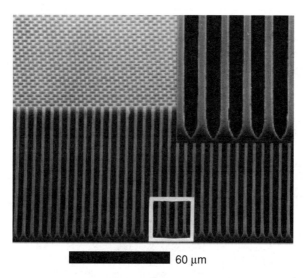

60 µm

FIGURE 2.12. Example of a regular array of macropores grown by Chao *et al.* [37]. p-Si (100) 13 Ω cm. Ethanolic HF electrolyte containing cetyltrimethylammonium chloride as a surfactant. Hexagonal array: pitch 5 µm. Pore depth 213 µm. Reproduced by permission of the Electrochemical Society.

Prepatterning has been made by pre-etching pyramidal etch pits arranged in a regular array [5,37,38]. This pre-etching was done by masking the Si surface with silicon oxide or nitride and conventional lithography, and exposing the masked surface to hot KOH solution. Nucleation of the macropores starts at the preetched pits. Perfectly regular growth can be achieved if the pitch of the nucleation array matches the average pore spacing in the conditions chosen for the anodization. If there is a significant mismatch between the prepatterning and the spontaneous pore spacing, the growth will be unstable. If the prepatterned pitch is too small, some pores will die so that the surviving ones can grow with their spontaneous spacing. If the prepatterned pitch is too large, extra pores will tend to appear between the prepatterned areas [38]. The matching condition, however, does not appear too stringent, and regular arrays of macropores have successfully been grown on p-Si. An example of such a growth is shown in Figure 2.12 [37]. Note that here the quality of pore growth was improved by the use of a cationic surfactant, an effect discussed in Section 2.4.3. Ordered arrays have been grown with a pitch of 5 µm and macropore aspect ratios of up to 100. In principle, the lower bound of the pitch size that can be realized is just limited by the size of the space charge, that is, by the lower bound of p-Si resistivity for which macropores can be grown. It is our prediction that any further increase of the band bending (by the adsorption of cationic species or the use of a suitable solvent), would be beneficial for extending this limit.

2.6. CONCLUSION

Though macropore growth on p-Si has appeared years after the corresponding studies started on n-Si, it seems to have reached a fair level of control and understanding. In general, crystallographic effects appear somewhat less marked for p-Si than for n-Si.

However, strongly anisotropic pore growth may be obtained in suitable non-aqueous solvents. Although macropores, in p-Si as well as in n-Si, do not exhibit spontaneous long-range ordering, they can be grown as long-range ordered arrays if the growth is initiated by prepatterning. Especially, the possibility to grow structures down to lower and lower resistivities may lead one to obtain smaller structures from p-Si than from n-Si. This opens the way to a variety of applications, from the manufacturing of micromechanical devices to the engineering of photonic-crystal materials.

REFERENCES

[1] R.L. Smith and S.D. Collins, J. Appl. Phys. **71**, R1–R22 (1992).
[2] L.T. Canham (Ed.), *Properties of Porous Silicon, EMIS Datareviews Series*, INSPEC, IEE, London, 1997.
[3] J.-N. Chazalviel, R.B. Wehrspohn, F. Ozanam and I. Solomon, MRS Symp. Proc. **452**, 403–414 (1997).
[4] R.B. Wehrspohn, F. Ozanam and J.-N. Chazalviel, J. Electrochem. Soc. **145**, 2958–2961 (1998).
[5] V. Lehmann and S. Rönnebeck, J. Electrochem. Soc. **146**, 2968–2975 (1999).
[6] E.K. Propst and P.A. Kohl, J. Electrochem. Soc. **141**, 1006–1013 (1994).
[7] M.M. Rieger and P.A. Kohl, J. Electrochem. Soc. **142**, 1490–1495 (1995).
[8] E.A. Ponomarev and C. Levy-Clément, Electrochemical Solid-State Lett. **1**, 42–45 (1998).
[9] E.A. Ponomarev and C. Levy-Clément, J. Porous Mater. **7**, 51–56 (2000).
[10] S. Lust and C. Levy-Clément, Phys. Status Solidi a **182**, 17–21 (2000).
[11] M. Christophersen, J. Carstensen, A. Feuerhake and H. Föll, Mater. Sci. Eng. B **69–70**, 194–198 (2000).
[12] C. Jäger, B. Finkenberger, W. Jäger, M. Christophersen, J. Carstensen and H. Föll, Mater. Sci. Eng. B **69–70**, 199–204 (2000).
[13] M. Christophersen, J. Carstensen and H. Föll, Phys. Status Solidi a **182**, 103–107 (2000).
[14] M. Christophersen, J. Carstensen, S. Rönnebeck, C. Jäger and H. Föll, J. Electrochem. Soc. **148**, E267–E275 (2001).
[15] A. Belaïdi, M. Safi, F. Ozanam, J.-N. Chazalviel and O. Gorochov, J. Electrochem. Soc. **146**, 2659–2664 (1999).
[16] J.-N. Chazalviel, M. Etman and F. Ozanam, J. Electroanal. Chem. **297**, 533–540 (1991).
[17] M. Etman, M. Neumann-Spallart, F. Ozanam and J.-N. Chazalviel, J. Electroanal. Chem. **301**, 259–265 (1991).
[18] V. Lehmann, J. Electrochem. Soc. **140**, 2836–2843 (1993).
[19] H.H. Hassan, J.L. Sculfort, M. Etman, F. Ozanam and J.-N. Chazalviel, J. Electroanal. Chem. **380**, 55–61 (1995).
[20] S. Cattarin, I. Frateur, M. Musiani and B. Tribollet, J. Electrochem. Soc. **147**, 3277–3282 (2000).
[21] J.E.A.M. Van den Meerakker and M.R.L. Mellier, J. Electrochem. Soc. **148**, G166–G171 (2001).
[22] H. Harada, T. Shirahashi, M. Nakamura, T. Ohwada, Y. Sasaki, S. Okuda and A. Hosono, Jpn. J. Appl. Phys. **40**, 4862–4863 (2001).
[23] J.-N. Chazalviel, F. Ozanam, N. Gabouze, S. Fellah and R.B. Wehrspohn, J. Electrochem. Soc. **149**, C511–C520 (2002).
[24] V. Lehmann and H. Föll, J. Electrochem. Soc. **137**, 653–659 (1990).
[25] J. Carstensen, M. Christophersen and H. Föll, Mater. Sci. Eng. B **69–70**, 23–28 (2000).
[26] W.W. Mullins and R.F. Sekerka, J. Appl. Phys. **35**, 444–451 (1964).
[27] Y. Kang and J. Jorné, J. Electrochem. Soc. **140**, 2258–2265 (1993).
[28] A. Valance, Phys. Rev. B **52**, 8323–8336 (1995).
[29] A. Valance, Phys. Rev. B **55**, 9706–9715 (1997).
[30] R.B. Wehrspohn, F. Ozanam and J.-N. Chazalviel, J. Electrochem. Soc. **146**, 3309–3314 (1999).
[31] J.-N. Chazalviel, R.B. Wehrspohn and F. Ozanam, Mater. Sci. Eng. B **69–70**, 1–10 (2000).
[32] I. Ronga, A. Bsiesy, F. Gaspard, R. Hérino, M. Ligeon, F. Muller and A. Halimaoui, J. Electrochem. Soc. **138**, 1403–1407 (1991).
[33] M. Christophersen, J. Carstensen and H. Föll, Phys. Status Solidi a **182**, 45–50 (2000).
[34] V. Lehmann and U. Gösele, Appl. Phys. Lett. **58**, 856–858 (1991).

[35] R. Memming and G. Schwandt, Surf. Sci. **4**, 109–124 (1966).
[36] E. Peiner and A. Schlachetzki, J. Electrochem. Soc. **139**, 552–557 (1992).
[37] K.J. Chao, S.C. Kao, C.M. Yang, M.S. Hseu and T.G. Tsai, Electrochem. Solid-State Lett. **3**, 489–492 (2000).
[38] A. Vyatkin, V. Starkov, V. Tzeitlin, H. Presting, J. Konle and U. König, J. Electrochem. Soc. **149**, G70–G76 (2002).

3

Highly Ordered Nanohole Arrays in Anodic Porous Alumina

Hideki Masuda

Department of Applied Chemistry, Tokyo Metropolitan University, 1-1 Minamiosawa,
Hachioji, Tokyo 192-03, Japan
Masuda-hideki@c.metro-u.ac.jp

3.1. INTRODUCTION

Anodic porous alumina, which is formed by the anodization of Al in appropriate acidic or basic electrolytic solutions [1–3], has recently attracted much interest as a starting material for the fabrication of several kinds of nanodevices due to its fine porous structure with high aspect ratio. The recent improvement in the degree of ordering of the anodic porous alumina [4–7] has increased the attractiveness of this material from the viewpoint of nanofabrication.

The geometrical structure of anodic porous alumina is schematically represented as a honeycomb structure consisting of a close-packed array of columnar alumina units called cells, each containing a central straight hole (Figure 3.1). The dimensions of the anodic porous alumina cells depend on the anodizing conditions [1,2]. The cell size, which is equivalent to the hole interval, is determined by the applied voltage used for the anodization; the cell size has a good linear relationship with the applied voltage [3]. The value of the constant of the cell size divided by the applied voltage is approximately 2.5 nm/V. The hole size is dependent on the electrolyte composition, temperature, period of anodization as well as applied voltage. The hole size is also controlled by the pore-widening treatment by dipping the porous alumina in an appropriate acid solution after the anodization. The cell size usually ranges from 10 to 500 nm and the hole size from 5 to 400 nm depending on the anodizing and post-anodizing conditions. The depth of the holes (thickness of the oxide films) has a good linear relationship with the period of the anodization.

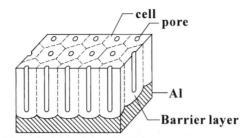

FIGURE 3.1. Schematic structure of anodic porous alumina.

3.2. NATURALLY OCCURRING LONG-RANGE ORDERING OF THE HOLE CONFIGURATION OF ANODIC ALUMINA

The ordering of the hole configuration of anodic porous alumina is essential for optimizing the performance of the obtained devices. The author and his coworkers have been studying the conditions for the naturally occurring long-range ordering of the hole configuration of anodic porous alumina in various kinds of acid electrolytes [4–7]. The long-range ordering of the hole configuration occurs under the appropriate anodizing conditions. The conditions of the long-range ordering are characterized by a long-period anodization under the appropriate constant anodizing voltages which are specific to the acid solution used for the anodization. The long-range ordering takes place at 25–27 V in sulfuric acid [5], 40 V in oxalic acid [4,6] and 195 V in phosphoric acid [7].

Figure 3.2 summarizes the SEM micrographs of the hole configuration with naturally occurring long-range ordering formed in these three acid electrolytes. These SEM micrographs were taken from the bottom side of the oxide layer called the barrier layer after the removal of the Al substrate with a saturated $HgCl_2$ solution. The barrier layer was removed with phosphoric acid solution, and then the pore-widening treatment was carried out using the same phosphoric acid solution. These treatments make it easy to observe the hole arrangement of the anodic porous alumina. In all of the SEM micrographs obtained in the three kinds of acids, the highly ordered hole configuration can be confirmed.

Figure 3.3 shows the dependence of the ordering of the hole configuration on anodizing period in sulfuric acid solution [5]. At the initial stage of the anodization, the holes develop randomly over the aluminium surface. During the growth of the oxide layer, the long-range ordering proceeds through the rearrangement of the hole configuration which changes gradually from a random configuration at the initial stage of anodization to a highly ordered configuration after the long-period anodization. In the hole configuration, characteristic patterns are observed in the array of holes as indicated by arrows (Figure 3.3a). These patterns correspond to the defect sites of the holes. As the ordering of the hole configuration proceeded, the number of these patterns decreased and an almost ideally ordered hole configuration was finally obtained over the sample [8–13].

From the large-area view (Figure 3.4), the ideally ordered area shows a domain structure at the boundary of which defects and imperfections are accumulated. During

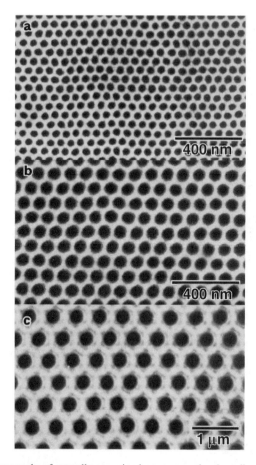

FIGURE 3.2. SEM micrographs of naturally occurring long-range ordered anodic porous alumina formed in three types of acid electrolytes: (a) sulfuric acid, (b) oxalic acid and (c) phosphoric acid.

the ordering of the hole configuration, such domains grow gradually, taking over the surrounding disordered holes, and attain an almost saturated size.

The manner of the self-ordering of hole configuration during anodization was almost the same in oxalic [4,6] and phosphoric acid [7] solutions except for the appropriate anodizing voltages. The saturated size of the ideally arranged domain is dependent on the size of the cells which compose the domain, i.e., it is largest in the phosphoric acid solution in which anodization is conducted under the highest applied voltage.

Concerning the naturally occurring self-ordering of the hole configuration of anodic porous alumina, similar results have been reported by other groups [8–12].

The detailed mechanism for explaining the dependence of the ordering on the applied voltage is not clear at the present stage. However, it appears that strain-free growth is feasible under the appropriate anodizing voltage.

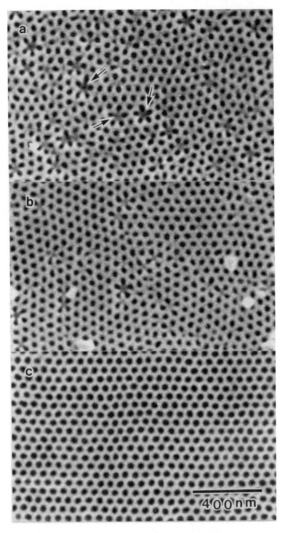

FIGURE 3.3. Dependence of the ordering of the hole arrangement of anodic porous alumina on the anodizing period in 0.5 M sulfuric acid solution at 25 V: (a) 9 minutes, (b) 36 minutes and (c) 710 minutes.

3.3. TWO-STEP ANODIZATION FOR ORDERED ARRAYS WITH STRAIGHT HOLES IN NATURALLY ORDERING PROCESSES

The degree of the ordering of the hole configuration at the surface of the anodic porous alumina is low because the holes develop randomly at the initial stage of the anodization. To improve the ordering of the surface side of the anodic porous alumina, two-step anodization is effectively adopted [6]. The process involves two separate anodization processes: the first anodization process consists of a long-period anodization to form the highly ordered hole configuration at the oxide/Al interface and the second

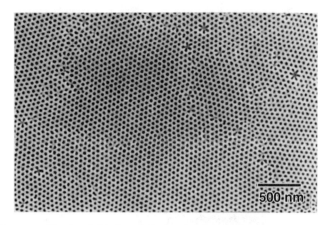

FIGURE 3.4. Low magnification SEM view of the long-range ordered anodic porous alumina formed in sulfuric acid at 25 V.

anodization is performed after the removal of the oxide formed in the first anodization step (Figure 3.5). After the removal of the oxide, an array of highly ordered dimples was formed on the Al, and these dimples can act as initiation sites for the hole development in the second anodization. This process generates an ordered hole array throughout the entire oxide layer. This process can also be applied for the preparation of a porous alumina mask used for several types of nanofabrication techniques in which the ordered straight through-holes are essential.

3.4. IDEALLY ORDERED HOLE ARRAY USING PRETEXTURING OF Al

In the case of the naturally occurring long-range ordering of the anodic porous alumina, the defect-free area forms the domain structure, and the size of the domain is limited to several micrometres. To form the ideally ordered single-domain structure over the sample, a new process using pretexturing of Al has been developed [14–16]. In this process, an array of shallow concaves is formed on Al by imprinting using a mould, and these concaves serve as initiation sites for the hole development at the initial stage of the anodization (Figure 3.6).

The master mould with a hexagonal array of the convexes was prepared by conventional electron beam (EB) lithography. The substrate used for the mould is a SiC single crystal wafer which has the mechanical strength and smoothness required for the mould.

The SiC mould is placed on the Al sheet, and pressing is conducted using an oil press at room temperature. After the imprinting, the pattern is fully transferred to the Al surface due to the sufficient plasticity of Al for mechanical moulding.

Figure 3.7 shows the SEM micrograph of the anodic porous alumina after the anodization of Al with a textured pattern [14]. In Figure 3.7, half of the sample (left-hand side) was pretextured using the mould. In the textured area, an ideally ordered hole configuration was formed, while the hole development was random in the untextured area.

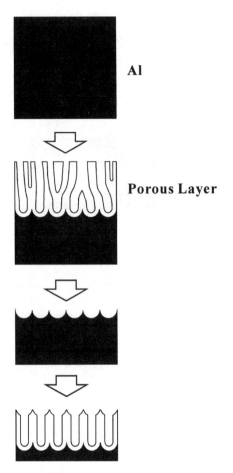

FIGURE 3.5. Schematic diagram of two-step anodization for the preparation of anodic porous alumina with straight holes.

Figure 3.8a shows the cross-sectional view of the sample of the single-domain structure. A highly ordered hole configuration and straight hole shape are confirmed throughout the film. The aspect ratio of the hole (hole length divided by the hole diameter) is approximately 150 in the sample of Figure 3.8. From the bottom side view shown in Figure 3.8b, it is confirmed that the ideally ordered hole configuration was maintained even at the bottom side of this high aspect ratio sample.

The lower limit of the dimensions of the fabrication of the hole array in the obtained structure is basically determined by the resolution of the patterning for the mould used for the imprinting. Figure 3.9 shows the SEM micrograph of the single-domain hole array with 63 nm interval which has the smallest period of holes at the present stage [15]. In the case of the sample in Figure 3.9, the hole size was 40 nm due to the post-etching treatment in phosphoric acid solution. The pore size was 15 nm in the case of the as-anodized sample without any pore-widening treatment.

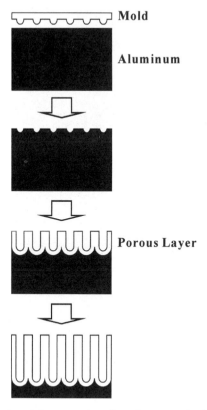

FIGURE 3.6. Schematic drawing of the preparation of anodic porous alumina with ideally ordered hole configuration using pretexturing process.

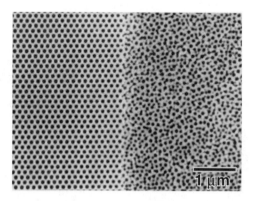

FIGURE 3.7. SEM micrograph of the surface of the anodic porous alumina using pretexturing process (left: prepatterned, right: no prepatterning).

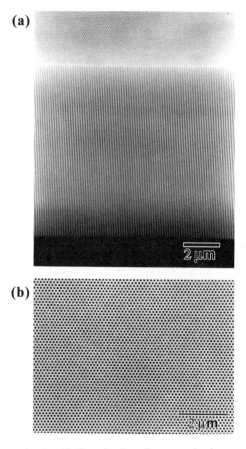

FIGURE 3.8. SEM micrographs of an ideally ordered anodic porous alumina: cross-sectional view (a) and bottom side view (b).

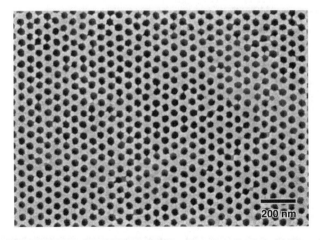

FIGURE 3.9. SEM micrograph of the ideally ordered anodic porous alumina with 63 nm hole interval.

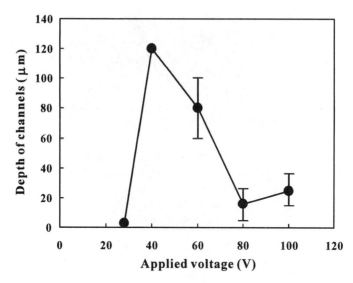

FIGURE 3.10. Dependence of the depth of holes with ideal ordering on the applied voltage for anodic porous alumina in oxalic acid.

Even in the case of anodization using the pretexuring of Al, appropriate anodizing conditions are essential for the fabrication of a hole array structure with a high aspect ratio [15,16]. Figure 3.10 shows the dependence of the depth of the ordered holes on the applied voltage for the anodization in oxalic acid solution [16]. Figure 3.10 shows that the most appropriate voltage for maintaining the ideal ordering is 40 V, and the ideal hole configuration with high aspect ratio cannot be maintained under other applied voltages. This voltage is in good accordance with the most appropriate voltage for the naturally occurring ordering in the oxalic acid. This result implies that the condition for the naturally occurring ordering is also important for maintaining the ideally ordered hole configuration induced by the pretexturing of Al.

Control of the hole development in anodic porous alumina by the use of a pretexturing process has also been reported by other groups. Liu *et al.* used the fast ion beam (FIB) technique for preparing the array of concaves on Al, which act as initiation sites for the hole development, and obtained the ideally ordered hole configuration of anodic porous alumina [17].

3.5. SELF-REPAIR OF THE HOLE CONFIGURATION IN ANODIC POROUS ALUMINA

One of the unique properties of anodic porous alumina is the self-repair of the hole configuration based on the self-compensation properties [18]. In the case of the anodization of pretextured Al using an imprinting process, even at the deficiency sites of the concaves, holes can be compensated automatically and an almost perfect arrangement of the hole configuration is generated. This is because that even at the deficiency sites, holes developed and then the closest packing of the cylindrical cells was recovered automatically.

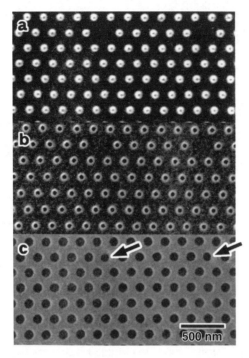

FIGURE 3.11. Self-repair of the hole configuration in anodic porous alumina: (a) SiC mould used for pretexturing, (b) pretextured Al and (c) anodic porous alumina with compensated holes at deficiency sites.

Figure 3.11 shows the results of the self-repair of the ordered pattern in anodic porous alumina [18]. In this experiment, the model pattern, in which some defects were introduced intensively, was used (Figure 3.11a). Figure 3.11b shows the SEM micrograph of the Al after the imprinting using a SiC mould with some defects in the pattern. After the anodization, the development of the holes could be observed even in the deficiency sites as shown by the arrows (Figure 3.11c). Using such a repaired structure as a template, moulds with a perfect arrangement of convexes for imprinting could be fabricated with metal.

This process is useful for the recovery of defects generated in the conventional lithographic processes and contributes to the fabrication of the defect-free fine pattern over a large area.

3.6. MODIFICATION OF THE SHAPE OF HOLE OPENING IN THE ANODIC POROUS ALUMINA

The shape of the cells of the anodic porous alumina can be determined by the mathematical expression known as the Voronoi tessellation. In the Voronoi tessellation, the pattern can be generated by a polygon, the boundary of which is a perpendicular bisector of two adjacent nuclei. In the naturally formed anodic porous alumina, the shape of the Voronoi cell is a hexagon with a triangular lattice. If the initiation nuclei sites can be laid out in other lattices, other cell shapes can be expected to form based

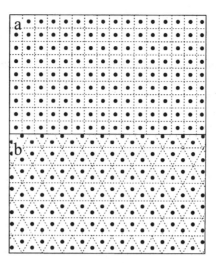

FIGURE 3.12. Formation of square and triangular cells based on Voronoi tessellation by controlling the layout of initiation site for the hole development.

on the Voronoi tessellation [19]. The shape of the cells is square in the square lattice and triangular in the graphite lattice, respectively (Figure 3.12). For the layout of the initiation site, the pretexturing process using a mould is successively used.

Figure 3.13 shows the SEM micrograph of the obtained anodic porous alumina with square and triangular openings [19]. The shape of the holes in the as-anodized

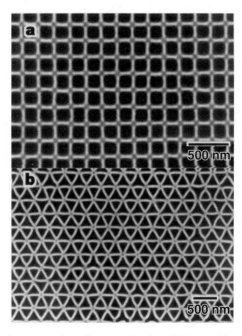

FIGURE 3.13. SEM micrographs of anodic porous alumina with square (a) and triangular holes (b).

anodic porous alumina was circular. However, the shape of the opening was changed from circular to square or triangular corresponding to the cell shape after the post-etching treatment in a phosphoric acid solution.

The most important concept of this process is that the shape of the obtained constituent units (cells, and cross section of holes) is determined by the arrangement of the nuclei in the two-dimensional space. This concept contributes to a higher resolution than that by the conventional pattering process where the pattern is formed through the painting of the figures by a stylus of limited size.

3.7. NANOFABRICATION BASED ON HIGHLY ORDERED ANODIC ALUMINA

3.7.1. 2D Photonic Crystals Using Anodic Porous Alumina

One of the promising application fields of the highly ordered hole array structure of the anodic porous alumina is two-dimensional (2D) photonic crystals (also see Chapter 7 by Wehrspohn and Schilling). The photonic crystals have a specially periodic refractive index with a lattice constant on the order of the wavelength of light, and show unique light propagation properties which can be used for the design of novel optoelectronic devices [20]. Works on the fabrication of 2D photonic crystals in the near infrared or visible wavelength are limited because of the difficulty in the fabrication of the ordered periodic structure with high aspect features. The geometrical structure of anodic porous alumina with an ordered array of air cylinders of a triangular lattice is one of typical 2D photonic crystals composed of air cylinders with a triangular lattice in an alumina matrix [21–23]. Photonic crystals based on anodic porous alumina are convenient to prepare and it is easy to control their dimensions.

The samples were cut in the specific directions in the air cylinder array, Γ–X, Γ–J (Figure 3.14), and optical measurement was carried out using a polarized light of E (E field paralleled to the air cylinders) and H (H field paralleled to air cylinders). Figure 3.15 shows the typical optical properties of the 2D photonic crystal prepared from the anodic porous alumina [21]. Distinct dips in the transmission spectra were observed for H polarization incident light in both Γ–X, and Γ–J directions, and an overlap wavelength region was observed. This dip in the transmission spectra was in good agreement with the calculation results for the hole array structure of anodic porous alumina. These results confirm that the anodic porous alumina with an ideally ordered hole configuration shows a photonic band gap in the visible wavelength region.

In addition to the ideally ordered anodic porous alumna obtained using the pretexturing process, naturally occurring ordered porous alumina also shows the photonic band gap corresponding to the hole period [23]. In this case, anodic porous alumina with the domain structure is interpreted as a polycrystalline type of photonic crystal.

3.7.2. Nanocomposite Using Highly Ordered Anodic Porous Alumina

There have been a large number of reports on the preparation of nanocomposite or nanocylinder structures using anodic porous alumina as a template [24–33]. The improvement of the hole configuration by naturally occurring long-range ordering of

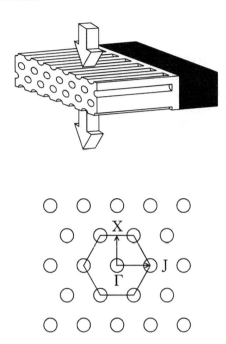

FIGURE 3.14. Top: Schematic diagram of the measurement configuration of 2D photonic band gap in anodic porous alumina. Bottom: Brioullinzone-free hexagonal porous alumina.

anodic porous alumina contributes to the advancement of the performance of the obtained nanodevices [34–41].

    One of the promising application fields of the highly ordered anodic porous alumina is for the preparation of the array of electron emitters in which uniform-sized carbon nanotubes are imbedded in the holes of anodic porous alumina [35–37]. An ordered hole

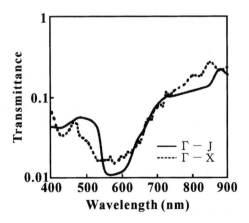

FIGURE 3.15. Transmission spectra of anodic porous alumina with 250 nm hole interval for H-polarized incident light.

array of anodic porous alumina was also used for CVD synthesis of nanocylindrical diamond [42].

Another application field of the long-range ordered anodic porous alumina is the fabrication of magnetic recording media [38–41] (see Chapter 8 by Nielsch *et al.*) Deposition of a ferromagnetic material such as Co or Ni into the micropores with high aspect ratio generates high-density perpendicular magnetic recording media. Although there have been large numbers of the reports on the magnetic recording media using anodic porous alumina [43,44], a highly ordered hole configuration contributes to the improvement of the S/N ratios of magnetic recording media using an anodic porous alumina matrix.

3.7.3. Two-Step Replication Process of the Hole Array Structure of Anodic Porous Alumina

Anodic porous alumina has several disadvantageous points in terms of its application in several areas: frangibility, low chemical stability, non-conductivity and so on. To overcome these disadvantageous points, and to expand the application fields of the anodic porous alumina, a new type of process, the so-called two-step replication process, has been developed [4,45–53]. This process consists of two main steps: the fabrication of a negative type of anodic porous alumina with a polymer (typically polymethylmethacrylate (PMMA)) and the subsequent reproduction of a positive type having an identical geometrical structure to starting anodic porous alumina (Figure 3.16). The important feature that distinguishes the present process from the usual one-step embedded process is that it permits the full replication of the hole array structure of anodic porous alumina and yields ordered hole array structures with desired materials.

For the preparation of the metal (Au, Ag, Pt, Ni, Pd, etc.) hole arrays, the electrochemical or electroless deposition of metals is adopted. Figure 3.17 shows the SEM micrograph of the typical hole array of Pt obtained by this process [4]. This process is also applied for the preparation of semiconductor hole array structures. Sol–gel deposition (TiO_2) [51] and electrochemical deposition (TiO_2, CdS) [52,53] are used for the injection of the semiconductors into the negative-type PMMA, and a semiconductor with ordered hole configuration is obtained.

This process has also been applied for the fabrication of nanohole arrays with other materials [54,55].

The obtained hole arrays are utilized as functional electrodes, sensors and photovoltaic cells. The hole array structure replicated with a high refractive index will also be useful as 2D phonic crystals in which a high contrast of refractive index is essential for obtaining the wide photonic band gap.

3.7.4. Fabrication of Nanodot and Nanohole Arrays Using Porous Alumina Masks

Anodic porous alumina with ordered straight through-holes can be applied to the mask for the fabrication of the nanodot or nanohole arrays [6,56–58]. After the anodization, the through-hole membrane is formed by removing the Al layer in a saturated $HgCl_2$ solution followed by the subsequent etching of the barrier layer in phosphoric acid

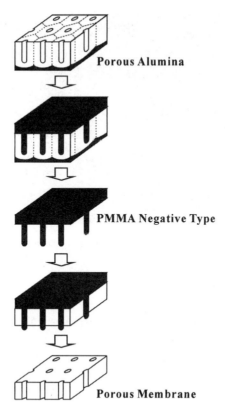

Porous Alumina

PMMA Negative Type

Porous Membrane

FIGURE 3.16. Two-step replication process using anodic porous alumina via PMMA negative.

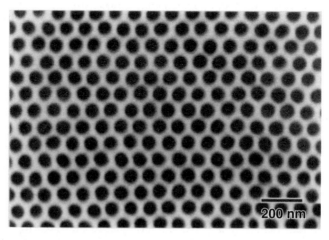

FIGURE 3.17. SEM micrograph of Pt hole array formed by two-step replication.

FIGURE 3.18. Nanodot array of Au on Si formed using anodic porous alumina mask with straight through-holes.

solution [6]. Two-step anodization is effective for the preparation of straight through-holes required for the mask applications in the case of naturally occurring anodic porous alumina [6].

The nanofabrication process using an anodic porous alumina mask is advantageous for the conventional lithographic process in the following: (1) a very fine pattern can be obtained over a large area, (2) the dimensions of the obtained structures can be controlled by changing the geometry of the anodic porous alumina mask and (3) the mask has holes with a high aspect ratio compared to the resist mask used in the conventional lithographic process.

Figure 3.18 shows the SEM micrograph of the metal (Au) nanodot array prepared on a Si substrate using an anodic porous alumina mask [6]. Au of 50 nm thickness was deposited through the anodic porous alumina mask using electron beam vacuum evaporation. The size and interval of the Au dots almost correspond to that of the alumina mask.

Making the characteristic feature of the alumina mask with a high aspect ratio, an array of multiple dots can also be fabricated based on the shadowing effect [56]. The sequential deposition of metals by changing the incidence angle yields an array of multiple metal dots on the substrate. The obtained array of multiple dots of nanometre dimension can be applied to the preparation of optical devices or model catalysts which requires an ordered array of multiple metal dots with nanometre dimensions.

The through-hole anodic porous alumina can also be used as a mask for the dry etching of several kinds of substrates [57–59]. The high resistance of the anodic porous alumina for the reactive etching plasma in addition to its high aspect ratio feature contributes to the fabrication of a hole array with high aspect ratio.

The use of the self-standing anodic porous alumina mask is simple and easy to perform on any type of substrates. However, the reproducibility and uniformity are insufficient because of the low adhesion of the mask to the substrates. To improve the adhesion of the mask to the substrates, the use of porous alumina prepared from the

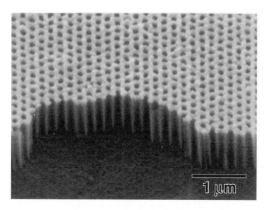

FIGURE 3.19. Anodic porous alumina mask on Si with an ideally ordered hole configuration obtained by anodization of pretextured Al on Si.

vacuum-evaporated Al was also used [9,35,58,60,61]. This process yields the mask preparation with high reproducibility and uniformity. An imprinting process using a mould is also applied to the Al on Si substrates, where the imprinting of Al is conducted using a SiC mould with hexagonally arranged convexes [58]. Figure 3.19 shows the SEM micrographs of the alumina mask prepared on a Si substrate using the imprinting process. The geometrical structure of the bottom part of the barrier layer on Si was different from that in bulk Al, that is, the shape of the barrier layer prepared from the Al/Si systems has voids beneath the pores [60]. This unique structure makes it easy to remove the barrier layer and generate the through-holes by the post-etching treatment in phosphoric acid solution.

3.8. CONCLUSION

Anodic porous alumina with highly ordered structures could be formed based on two types of ordering processes: naturally occurring ordering under the appropriate anodizing conditions and anodization using pretextured Al. The fabrication based on the naturally occurring long-range ordering is simple and useful for the ordered hole array configuration with large area. Anodic porous alumina with an ideally ordered hole array obtained by the pretexturing process is useful for applications in which a strictly ordered single-domain hole configuration is required, such as optical devices or patterned recording media. Both types of ordered anodic porous alumina are promising as starting structures for the fabrication of a wide variety of functional devices with nanometre dimensions.

REFERENCES

[1] F. Keller, M. Hunter and D.L. Robinson, J. Electrochem. Soc. 100, 411 (1953).
[2] J.P.O'Sullivan and G.C. Wood, Proc. R. Soc. Lond., Ser. A 317, 511 (1970).

[3] K. Ebihara, H. Takahashi and M. Nagayama, J. Surf. Fin. Soc. Jpn. **34**, 548 (1983).

[4] H. Masuda and K. Fukuda, Science **268**, 1466 (1995).

[5] H. Masuda, F. Hasegawa and S. Ono, J. Electrochem. Soc. **144**, L127 (1997).

[6] H. Masuda and M. Satoh, Jpn. J. Appl. Phys. **35**, L126 (1996).

[7] H. Masuda, K. Yada and A. Osaka, Jpn. J. Appl. Phys. **37**, L1340 (1998).

[8] S. Shingubara, O. Okino, Y. Sayama, H. Sakaue and T. Takahagi, Jpn. J. Appl. Phys. **36**, 7791 (1997).

[9] S. Shingubara, O. Okino, Y. Sayama, H. Sakaue and T. Takahagi, Solid-State Electron. **43**, 1143 (1999).

[10] O. Jessensky, F. Muller and U. Gosele, Appl. Phys. Lett. **72**, 1173 (1998).

[11] A.P. Li, F. Muller, A. Birner, K. Nielsch and U. Gosele, J. Appl. Phys. **84**, 6023 (1998).

[12] F. Li, L. Zhang and R.M. Metzger, Chem. Mater. **10**, 2470 (1998).

[13] L. Zhang, H.S. Cho, F. Li, R.M. Metzger and W.D. Doyle, J. Mater. Sci. Lett. **17**, 291 (1998).

[14] H. Masuda, H. Yamada, M. Satoh, H. Asoh, M. Nakao and T. Tamamura, Appl. Phys. Lett. **71**, 2770 (1997).

[15] H. Asoh, K. Nishio, M. Nakao, A. Yokoo, T. Tamamura and H. Masuda, J. Vac. Sci. Tech. B **19**, 569 (2001).

[16] H. Asoh, K. Nishio, M. Nakao, T. Tamamura and H. Masuda, J. Electochem. Soc. **148**, B152 (2001).

[17] C.Y. Liu, A. Datta and Y.L. Wang, Appl. Phys. Lett. **78**, 120 (2001).

[18] H. Masuda, M. Yotsuya, M. Asano, K. Nishio, M. Nakao, A. Yokoo and T. Tamamura, Appl. Phys. Lett. **78**, 826 (2001).

[19] H. Masuda, H. Asoh, M. Watanabe, K. Nishio, M. Nakao and T. Tamamura, Adv. Mater. **13**, 189 (2001).

[20] J.D. Jannopoulos, R.D. Meade and J.N. Winn, *Photonic Crystals*, Princeton University Press, Princeton, 1995.

[21] H. Masuda, M. Ohya, H. Asoh, M. Nakao, M. Nohtomi and T. Tamamura, Jpn. J. Appl. Phys. **38**, L1403 (1999).

[22] H. Masuda, M. Ohya, K. Nishio, H. Asoh, M. Nakao, M. Nohtomi, A. Yokoo and T. Tamamura, Jpn. J. Appl. Phys. **39**, L1039 (2000).

[23] H. Masuda, M. Ohya, H. Asoh and K. Nishio, Jpn. J. Appl. Phys. **40**, L1217 (2001).

[24] C.R. Martin, Science **266**, 1961 (1994).

[25] J.C. Hulteen and C.R. Martin, J. Mater. Chem. **7**, 1075 (1997).

[26] G. Che, B.B. Lakshmi, E.R. Fisher and C.R. Martin, Nature **393**, 346 (1998).

[27] C.K. Preston and M. Moskovits, J. Phys. Chem. **92**, 2957 (1988).

[28] C.K. Preston and M. Moskovits, J. Phys. Chem. **97**, 8495 (1993).

[29] D. Routkevitch, T. Bigioni, M. Moskovits and J.M. Xu, J. Phys. Chem. **100**, 14037 (1996).

[30] S. Kawai and R. Ueda, J. Electrochem. Soc. **122**, 32 (1975).

[31] C.G. Granqvist, A. Anderson and O. Hunderi, Appl. Phys. Lett. **35**, 268 (1979).

[32] C.A. Huber, T.E. Huber, M. Sadoqi, J.A. Lubin, S. Manalis and C.B. Prater, Science **263**, 800 (1994).

[33] T. Kyotani, L.-F. Tasi and A. Tomita, Chem. Mater. **8**, 2109 (1996).

[34] J. Zhang, L.D. Zhang, X.F. Wang, C.H. Liang, X.S. Peng and Y.W. Wang, J. Chem. Phys. **115**, 5714 (2001).

[35] T. Iwasaki, T. Motoi and T. Den, Appl. Phys. Lett. **75**, 2044 (1999).

[36] J.S. Suh and J.S. Lee, Appl. Phys. Lett. **75**, 2047 (1999).

[37] S.-H. Jeong, H.-Y. Hwang, K.-H. Lee and Y. Jeong, Appl. Phys. Lett. **78**, 2052 (2001).

[38] F. Li, R. Metzger and W.D. Doyle, IEEE Trans. Mag. **33**, 3715 (1997).

[39] S.G. Yang, H. Zhu, G. Ni, D.L. Yu, S.L. Tang and Y.W. Du, J. Phys. D **33**, 2388 (2000).

[40] M. Zheng, L. Menon, H. Zeng, Y. Liu, S. Bandyopadhyyay, R.D. Kirby and D. J. Sellmyer, Phys. Rev. B **62**, 12282 (2000).

[41] K. Nielsch, R.B. Wehrspohn, J. Barthel, J. Kirschner, U. Gosele, S.F. Fischer and H. Kronmuller, Appl. Phys. Lett. **79**, 1360 (2001).

[42] H. Masuda, T. Yanagishita, K. Yasui, K. Nishio, I. Yagi, T. Rao and A. Fujishima, Adv. Mater. **13**, 247 (2001).

[43] T.J. Cheng, J. Jorne and J.-S. Gau, J. Electrochem. Soc. **137**, 93 (1990).

[44] H. Daimon, O. Kitakami, O. Inagoya, A. Sakemoto and K. Mizushima, Jpn. J. Appl. Phys. **29**, 1675 (1990).

[45] H. Masuda H. Tanaka and N. Baba, Chem. Lett. **621** (1990).

[46] H. Masuda, H. Tanaka and N. Baba, Bull. Chem. Soc. Jpn. **66**, 305 (1993).

[47] H. Masuda, K. Nishio and N. Baba, Thin Solid Films **223**, 1 (1993).
[48] H. Masuda, T. Mizuno, N. Baba and T. Ohmori, J. Electroanal. Chem. **368**, 333. (1994).
[49] H. Masuda and K. Fukuda, J. Electroanal. Chem. **473**, 240 (1999).
[50] T. Ohmori, T. Kimura and H. Masuda, J. Electrochem. Soc. **144**, 1286 (1997).
[51] H. Masuda, K. Nishio and N. Baba, Jpn. J. Appl. Phys. **31**, L1775 (1992).
[52] P. Hoyer, N. Baba and H. Masuda, Appl. Phys. Lett. **66**, 2700 (1995).
[53] P. Hoyer and H. Masuda, J. Mater. Sci. Lett. **15**, 1228 (1996).
[54] Y. Lei, C.H. Liang, Y.C. Wu, L.D. Zhang and Y.Q. Mao, J. Vac. Sci. Tech. B **19**, 1109 (2001).
[55] K. Jiang, Y. Wang, J. Dong, L. Gui and Y. Tang, Langmuir **17**, 3635 (2001).
[56] H. Masuda, K. Yasui and K. Nisho, Adv. Mater. **12**, 1031 (2000).
[57] M. Nakao, S. Oku, T. Tamamura, K. Yasui and H. Masuda, Jpn. J. Appl. Phys. **38**, 1052 (1999).
[58] H. Masuda, K. Yasui, Y. Sakamoto, M. Nakao, T. Tamamura and K. Nishio, Jpn. J. Appl. Phys. **40**, L1267 (2001).
[59] Y. Kanamori, K. Hane, H. Sai and H. Yugami, Appl. Phys. Lett. **78**, 142 (2001).
[60] D. Crouse, Y.-H. Lo, A.E. Miller and M. Crouse, Appl. Phys. Lett. **76**, 49 (2000).
[61] J.H. Wu, X.L. Wu, N. Tang, Y.F. Mei, X.M. Bao, Appl. Phys. A **72**, 735 (2001).

4

The Way to Uniformity in Porous III–V Compounds via Self-Organization and Lithography Patterning

S. Langa[1,2], J. Carstensen[1], M. Christophersen[1], H. Föll[1] and I.M. Tiginyanu[2]

[1]*Materials Science Department, Faculty of Engineering, Christian-Albrechts-University, Kaiserstraße 2, D-24143, Kiel, Germany*
hf@tf.uni-kiel.de
[2]*Laboratory of Low Dimensional Semiconductor Structures, Technical University of Moldova, St. cel Mare 168, MD-2004, Chisinau, Moldova*

4.1. INTRODUCTION

4.1.1. State of the Art in Research

Electrochemical etching techniques offer wide possibilities for modifying the surface morphology of semiconductors. Depending on the substrate type, electrolyte, illumination, etc., anodic etching may lead to both the electropolishing and the formation of pores. The factors determining the morphology of electrochemically etched III–V compounds have been reviewed by Gomes and Goossens as well as by Notten *et al.* [1,2]. Under anodic conditions the p- and n-type materials were found to behave quite differently. For instance, in the case of p-GaAs substrates in acidic solutions, the current in the anodic direction rises rapidly to very high values causing electropolishing of the material [3]. At the same time the n-GaAs/aqueous electrolyte interface shows diode characteristics. Under forward bias a cathodic current is observed, while under reverse bias the current measured in the dark is very low. In the case of n-type materials under anodic bias, the current limiting factor seems to be the space-charge layer at the semiconductor surface, i.e., the electron transfer at the interface is limited by tunnelling through the depletion layer or by carriers that overcome the barrier by thermal activation. As soon as a critical potential (the so-called pitting potential) is reached a steep current increase occurs leading to the formation of pits on the surface [4]. Current limitation via oxide film formation may also play a role although this issue has not been studied systematically so far [5].

Depending on the surface orientation and etching conditions, different types of porous morphologies have been observed in III–V compounds subjected to anodization, e.g., crystallographically oriented (CO) pores, current-line-oriented pores, tetrahedron-like pores, "catacomb"-like pores, etc. Electrochemical etching of a (111)A-oriented n-InP surface in a HCl/H_2O solution can lead to a top layer with a pillar structure characterized by quasi-isolated columns perpendicular to the initial surface [6]. A similar feature consisting of triangular-prism-shaped columns has also been revealed on (111)A-oriented n-GaP surfaces electrochemically etched in a solution of HF in ethanol [7]. In contrast, porous layers on n-GaP and n-GaAs surfaces with a (100) orientation show a cellular structure after etching in H_2SO_4/H_2O solution [7,8]. Moreover, in n-type (100)-oriented GaP after the initial pitting of the surface, further etching proceeds in directions both perpendicular and parallel to the surface. Since the extension of the porous structure occurs underneath the surface, this is termed a "catacomb"-like porosity [9].

Porous etching in n-type III–V compounds is considered to start at defect sites, since the increased field strength and the high density of localized energy levels at the sites involved facilitate the hole generation necessary for the material dissolution. Further development of the porous structure seems to be strongly determined by different stabilities against the dissolution of distinct crystallographic planes. In n-GaAs, for instance, the preferential chemical attack was found to take place along the ⟨111⟩ directions [10,11]. Preferential alignment of pores along the ⟨111⟩ directions was also observed in electrochemically etched n-InP [12] and n-GaP [13]. The formation of triangular-prism-shaped pores in anodically etched (111)-oriented n-GaP and n-InP appears also to be related to the different chemical reactivities of the crystallographic planes.

A severe general problem of pore formation in III–V semiconductors is the homogeneity of the porous structures, reflecting the strong tendency for inhomogeneous primary nucleation. Only some attempts to prepare quasi-uniform porous structures in III–V materials have been undertaken so far. As mentioned above, Takizava et al. fabricated a pillar-like porous structure on (111)A-oriented InP using electron-beam lithography, but the obtained structures are not so far uniform as those obtained by means of lithography in Si [14]. Recently Tiginyanu et al. [9] demonstrated that a 5 MeV Kr^+ implantation in n-GaP substrates allows us to control the surface defect density and thus the nucleation, irrespective of its initial value determined by the crystal growth.

4.1.2. Why Porous III–V Materials?

Although the number of publications devoted to porous III–V compounds is much smaller than that devoted to porous Si, some important properties of porous III–V structures have been reported recently. In particular, the following new findings merit attention:

- A sharp increase in intensity of the near-band-edge photoluminescence (PL) in anodically etched GaP along with the emergence of blue and ultraviolet luminescence [15–17].
- A strongly enhanced photo response was observed during pore formation in n-GaP electrodes by anodic etching in sulfuric acid solution [7,18].
- Evidence has been found for birefringence in porous InP at wavelengths suitable for optical communication systems [19].
- Porosity-induced modification of the phonon spectrum was observed in GaP, GaAs and InP [20,21].
- A very efficient optical second harmonic generation (SHG) was observed in porous GaP membranes [22].

Thus, porous compound semiconductors could offer potential advantages for device applications.

Besides the inherent possibilities in changing the chemical composition, the shift from elementary to compound semiconductors entails a major crystallographic modification. Although the overall tetrahedral sp^3 bonding between atoms is kept, the centrosymmetrical lattice of the column IV element of diamond type becomes a non-centrosymmetrical lattice of sphalerite type for the derived IB–VII, II–VI and III–V binary compounds. This opens the way for new physical properties specific to the accentricity of these polar materials. For instance, we observed surface-related vibrational modes in the spontaneous Raman scattering (RS) spectra of porous III–V compounds [20,21]. These modes of the Fröhlich character are predicted to occur whenever the wavelength of the incident radiation becomes greater than the average size of the crystallites. In connection with the strong tendencies to further miniaturization in modern electronics, it is obvious that the properties of many elements in optoelectronic circuits are governed by surface excitations. The investigation of surface vibrations is crucial for understanding absorption–desorption, diffusion, stability of nanostructures, physics of phase transitions in reduced dimensionality, etc.

In full agreement with theoretical predictions, the Fröhlich mode in porous GaP was found to exhibit a downward frequency shift with increasing porosity and/or dielectric constant of the surrounding medium [20,21,23]. This behaviour hints at new possibilities for optical phonon engineering in polar materials. Since the frequency of a Fröhlich mode can be changed in a controllable manner within the frequency gap of the bulk optical phonons (TO-LO), one can propose a completely new design of phonon-assisted optoelectronic devices. In quantum well structure devices [24], for instance, the wells can be made porous. This will enable the use of the Fröhlich-vibration-assisted tunnelling as an operating principle, taking the advantage of the possibility to control this process by optical means.

Another important property inherent to accentricity is the optical SHG. Since porous structures represent heterogeneous materials characterized by a pronounced local field enhancement, the efficient SHG observed in porous GaP [22] is a subject of special interest from both the fundamental and practical points of view. Further progress in this field will obviously depend on the possibility to manufacture high-quality optically homogeneous porous layers and membranes.

Also, taking into account the wide electronic band gap of III–V compounds and the possibility to introduce quasi-uniform self-arranged distributed pores with sizes down to a few tens of nanometres, one can consider porous III–V compounds as promising materials for different photonic applications.

4.2. ASPECTS OF CHEMISTRY AND ELECTROCHEMISTRY OF SEMICONDUCTORS

4.2.1. Looking at Metals and Semiconductors from the Electrochemical Point of View

Electrochemical etching is accompanied by transfer of free charge carriers through the semiconductor/electrolyte interface (SEI) and the external circuit, while simple chemical dissolution requires a reactive electrolyte and no charge exchange through the external circuit is needed [25].

From the electrochemical point of view, semiconductor electrodes in contact with electrolytes behave very differently in comparison with metal ones [26]. This is based on some characteristic properties of semiconductors when they come in contact with electrolytes. The first characteristic feature is that the energy of the conduction and valence bands of the semiconductor at the interface is nearly fixed, i.e., they only move slightly relative to the energy levels of the ions in the electrolyte, since most of the applied potential drops in the semiconductor. This means that if the band edges are favourable for electron transfer to or from a redox couple, the transfer will occur independently of the applied voltage. The second special feature of semiconductor electrodes in redox reactions is the ability to control the band (valence or conduction) in which the reacting carrier will appear.

These unique properties of semiconductor electrodes allow us to measure the characteristics of ions in solutions and to define the steps of an electrochemical reaction with facilities not attainable with metal electrodes. Thus, semiconductors can allow us for a better understanding and control of the processes which play a major role in modern electronics [27,28] and also the prediction of new features which may be used for the fabrication of novel semiconductor structures.

4.2.2. Chemical Etching of III–V Semiconductors

During the chemical etching process the bonds between the atoms are broken and new bonds are formed with reactive molecules in the electrolyte. In order to etch chemically a III–V compound, it is necessary to have a strong reactive medium capable to break the III–V bonds and consequently to saturate the resulting dangling bonds. Examples of reactive molecules frequently used to etch chemically III–V compounds are H_2O_2, Br_2, etc. [29]. The chemical reaction is usually divided in some slow and quick reaction steps. The slowest step is the rate determining one. For example, during the chemical reaction of GaAs in H_2O_2 solutions, the breaking of the Ga–As bond (followed by the formation of GaOH and AsOH bonds) is the rate-determining step. The subsequent steps are much faster than the first one; therefore, they do not contribute much to the overall rate of the chemical reaction.

4.2.3. Electrochemical Etching

Electrochemical etching can be divided into an electrochemical process and a subsequent purely chemical etching.

In order to perform an anodic etching process, it is necessary to connect the semiconductor electrode to an external power source. The bonds can eventually be broken by applying a sufficiently positive voltage and therefore no strong reactive molecules are necessary to initiate the reaction. The resulting dangling bonds can react with the reducing molecules in the electrolyte. If the compound formed (atoms of the solid + electrolyte molecules) is soluble in the etching medium, it dissolves chemically and etching of the electrode occurs [30], otherwise on the surface of the electrode a thin oxide layer is formed which hinders the electrochemical attack to proceed. For this reason, the electrolytes should contain two main components: reducing molecules and active oxide dissolving species.

It is generally assumed that bond breaking is caused by accumulation of holes at the SEI. During anodic etching the holes can be supplied by applying a sufficiently positive (anodic) potential to the electrode by means of the so-called avalanche breakdown mechanism or by illuminating the semiconductor with photons having energies higher than the semiconductor band gap and thus generating electron–hole pairs.

Note that during anodic etching two possibilities for in situ illumination are possible: front-side and back-side illumination. Using front-side illumination, the SEI is illuminated directly by the light source and holes are produced close to where they are consumed. Uniform illumination leads to a uniform hole generation which in III–V compounds can lead to a uniform dissolution (electropolishing), while in Si still pronounced pores may be formed [31–33].

When the so-called back-side illumination is used [34], the back side of the wafer is illuminated and consequently the holes are generated far away from the region where the chemical reaction should take place. Therefore, the holes have to diffuse from the back to the front side of the wafer. In order for the back-side illumination process to be effective, the diffusion length of the holes must be high enough to allow at least some of the back-side-generated holes to reach the SEI. Back-side illumination is extensively used for pore formation processes in Si which has diffusion lengths for the holes in the range of some hundreds of micrometres.

Unfortunately, the back-side illumination is not applicable for III–V compounds. The reason is relatively simple. III–V compounds have small diffusion lengths for the holes and electrons which are in the range of some tens of nanometres. Taking into account that standard semiconductor wafers have a thickness in the range of 500 μm, it becomes clear that back-side illumination is not practicable for III–V compounds because the back-side-generated holes will recombine in the bulk of the semiconductor.

4.2.4. Breakdown Mechanism

As already mentioned, applying a positive potential to a semiconductor wafer will cause the drift of the positive charge carriers towards the SEI. Using n-type samples in the dark (no back- or front-side illumination), the number of holes (anodic current) reaching the SEI will be small due to the fact that holes in n-type semiconductors are

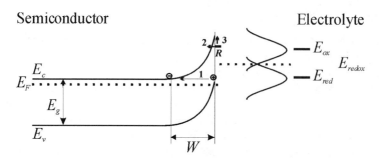

FIGURE 4.1. The band diagram of the SEI explaining the breakdown mechanism when a positive voltage is applied on the sample.

minority carriers. However, the anodic current remains low only for a small range of the electrode potential. At higher potentials a steep current rise is usually observed. In this case, the anodic current passing through the SEI is due to the tunnelling of electrons through the space-charge region coupled with avalanche processes. Tunnelling can occur if the band bending is larger than the band gap of the semiconductor. For moderately doped III–V compounds the thickness of the space-charge layer is several hundreds of angstroms in width and the number of holes produced by tunnelling is in general very low. Therefore, the steep increase in anodic currents at high voltages is assumed to be related to a local avalanche mechanism. The avalanche is initiated by a small number of tunnelled electrons (Figure 4.1, process 1) from the valence to the conduction band. If the field strength is high enough, the new created electrons can gain such a high energy that on their path from the SEI towards the bulk will generate new electron–hole pairs by means of the avalanche mechanism.

It should be noted that the oxidation intermediates (R in Figure 4.1) produced by the initial tunnelling process (process 1, Figure 4.1) can be viewed as surface states with an energy level above the valence band edge [11,35]. These intermediates are highly reactive and can be still oxidized by injection of an electron into the conduction band by thermal excitations (process 3, Figure 4.1) or by tunnelling (process 2, Figure 4.1).

Usually, a localized avalanche mechanism is assumed to start at surface defects, dislocations, etc. New electron–hole pairs are generated as soon as breakdown starts and thus etching around the defect will occur. The etching process induces new surface defects, e.g., pits, and the avalanche breakdown will occur at lower externally applied potentials as before [36]. Thus, the presence of defects determines the avalanche break-down and consequently the anodic current—in contrast to Si, where theoretical limits for perfect crystals can be reached for breakdown.

The holes generated by the avalanche mechanism will drift towards the SEI and the semiconductor will dissolve. It is generally accepted that in order to dissolve a III–V couple, for example, one unit of Ga–As, six holes are necessary [37]:

$$GaAs + 6h^+ \longrightarrow Ga^{3+} + As^{3+}. \tag{4.1}$$

The holes compensate for the six incomplete active bonds which can be saturated by E_{red} (Figure 4.1) from the electrolyte.

It is important to note that avalanche breakdown is not so evident for electrolytes with neutral pH values. In this case any initial avalanche event, localized at one of the defects, is immediately followed by oxide formation. The local resistance then is increased and etching at the defect stops. However, the local avalanche breakdown will always stop after a short period of time in any kind of electrolytes, not only in neutral pH solutions. The electrolyte and/or the semiconductor will quench the avalanche breakthrough due to diffusion losses or nonlinear ohmic effects.

4.2.5. Anisotropy of Dissolution in III–V Compounds

Anisotropic etching is of great importance for 2D and 3D semiconductor structuring. Anisotropy is mainly caused by the differences in etching rates of low-indexed crystal facets. The reasons for this behaviour are not yet completely understood.

The usually investigated low-indexed crystal faces are (111), (110) and (100). One of the reasons for the observed dissolution differences could be the density of atoms in these planes. However, this is not the only factor determining the anisotropy. In III–V compounds an additional factor contributing to anisotropy is related to the different chemical properties of the atoms from the third and fifth groups of the periodic table of the elements. In $\langle 111 \rangle$ directions, atomic planes are occupied alternatively by atoms from the third and fifth groups forming a double layer. The facets (111), $(1\bar{1}\bar{1})$, $(\bar{1}\bar{1}1)$, $(\bar{1}1\bar{1})$ are usually called $\{111\}$A, while $(\bar{1}\bar{1}\bar{1})$, $(\bar{1}11)$, $(11\bar{1})$, $(1\bar{1}1)$ are called $\{111\}$B. In nearly all oxidizing electrolytes $\{111\}$A facets show the slowest rate of dissolution [38].

The direction of pore growth in III–V compounds is also strongly influenced by the difference in the dissolution rate of the $\{111\}$A and $\{111\}$B facets. For example in Figure 4.2a, the schematic nucleation sequence of pores on (100)-oriented wafers is presented. The first step of the nucleation leads to the formation of pyramid-like pits exposing both types of $\{111\}$ planes, e.g., A and B (Figure 4.2a). As the etching process goes on, the difference in the dissolution rates of the two types of $\{111\}$ planes becomes more and more pronounced. Due to the fact that B planes are more easily etched, the initial pyramids begin to stretch along the directions marked as X (Figure 4.2b). In the same time the B planes rotate around the Z axis (Figure 4.2b) until an A-type plane [39] is reached which is more stable against dissolution. When this state is achieved, only A-type planes are exposed along the whole-elongated cavity. The pits have two sharp tips oriented along $\langle 111 \rangle$ directions (Figure 4.2c). Taking into account that the holes necessary for dissolution are created by means of the avalanche mechanism, it can be expected that at the tips of the pits where due to the strong curvature, the electric field strength is higher [40], more holes are generated and thus high dissolution rates are reached. Thus the tips of the pits will move along the $\langle 111 \rangle$ directions forming the pores.

The pores formed according to this mechanism will mainly expose the $\{112\}$ planes as walls (Figure 4.2d) [10]. However, pores exposing $\{110\}$ planes have also been reported [11].

Figure 4.3 illustrates a SEM picture from an anodized (100)-oriented GaAs sample showing the nucleation of pores according to the mechanism explained schematically in Figure 4.2.

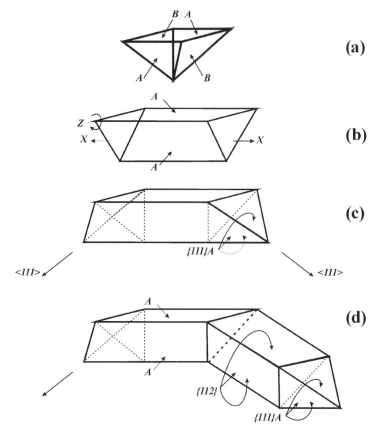

FIGURE 4.2. A schematic presentation of the nucleation of crystallographically oriented pores in III–V compounds.

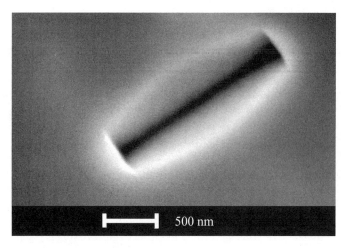

FIGURE 4.3. Nucleation of crystallographically oriented pores in (100)-oriented GaAs exposing {111}A planes as the most stable against dissolution.

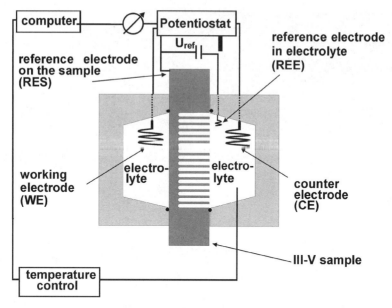

FIGURE 4.4. The experimental set-up used for anodization of III–V compounds.

4.3. PORE MORPHOLOGIES OBSERVED IN III–V COMPOUNDS

4.3.1. Experimental Set-Up

The anodization was carried out in a Teflon electrochemical double cell presented in Figure 4.4. A four electrode configuration was used: a Pt reference electrode in the electrolyte (RE), a Pt sense electrode on the sample (SE), a Pt counter electrode (CE) and a Pt working electrode (WE). The electrodes were connected to a specially manufactured potentiostat/galvanostat.

The electrolyte is pumped continuously through both parts of the double cell by means of peristaltic pumps. The equipment used in the experiments was computer controlled. The area of the sample exposed to the electrolyte was 0.2 cm^2.

4.3.2. Crystallographically Oriented Pores

In order to study the morphology of anodized samples fresh cleavages were prepared and analyzed using a scanning electron microscope (SEM). Figure 4.5 (a, c, d—GaAs; b—InP) shows cross-sectional SEM images taken from an n-type (100)-oriented sample anodized at $j = 4$ mA/cm^2 in 5% HCl.

An angle of 108° is found between the directions of pore growth (marked by arrows in Figures 4.5a and b). Thus, the two directions of pore growth can be identified as [111] and [1$\bar{1}\bar{1}$]. The intersection of these [111]- and [1$\bar{1}\bar{1}$]-oriented pores is most likely the reason for the formation of the visible terraces after the cleavage in the InP sample (Figure 4.5b). Apart from that, the analysis of the SEM micrographs presented in Figures 4.5a and b show the sets of pores oriented along [1$\bar{1}$1] and [11$\bar{1}$] crystallographic directions.

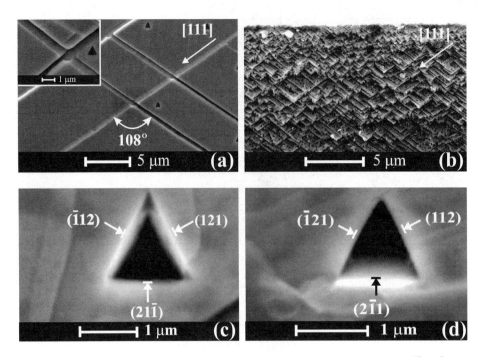

FIGURE 4.5. (a and b) Crystallographically oriented pores in (100)-oriented GaAs ($n = 10^{17} \text{cm}^{-3}$) and (100)-oriented InP ($n = 3 \times 10^{17} \text{ cm}^{-3}$); (c and d) the pores grow in $\langle 111 \rangle$ directions exposing the $\{112\}$ planes.

The pores show a triangular cross section and the pore walls are formed by $\{112\}$ planes. The formation of triangular pores introduces a second anisotropy of the electrochemical etching (besides the preferred directions of pore growth along $\langle 111 \rangle$) and establishes $\{112\}$ planes as particularly stable against dissolution.

The $\langle 111 \rangle$-oriented pores nucleating from a (100)-oriented surface can form a 3D structure if their intersection is assured. Figure 4.5a illustrates an example of the intersection of two pores oriented along [111] and [1$\bar{1}\bar{1}$] directions. Somewhat surprisingly, the intersection has no influence on the pore shape, the size and the direction of subsequent propagation. This important finding demonstrates that anodic etching may be a suitable and unique tool for the production of 3D micro- and nano-structured III–V compounds, e.g., for photonic crystals applications. Nevertheless, in addition to pore interconnection, a three-dimensional photonic pore crystal requires a very high level of uniformity [41]. Since the dissolution is stimulated by defects [9,42] and/or illumination [43], it is expected that a predefined nucleation and thus a controlled intersection of the pores can be provided by conventional lithographic means followed by a conventional chemical treatment to generate small pits (defects) or by the proper front-side illumination of the sample at the beginning of the anodization.

The crossing of pores is somewhat unexpected because according to some existing models [34,44], the formation of pores in n-type semiconductors is a self-adjusted process controlled by the distribution of the electric field at the semiconductor–electrolyte

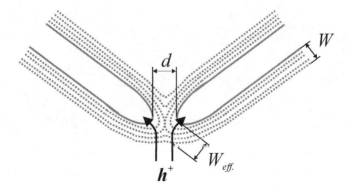

FIGURE 4.6. Schematic representation of the potential distribution around two pores which will intersect: W, space-charge layer width; $d < 2W$, the actual distance between the pore tips; the arrows mark the path of the holes h^+ generated by breakdown mechanism.

interface. The pores may branch and form porous domains but both individual pores and domains should be separated by walls with characteristic dimensions of twice the thickness of the surface depletion layer and cannot intersect [7]. These has been explained supposing that when a wall becomes too thin (smaller than twice the space-charge region), it can no longer support a field perpendicular to the surface which is sufficiently high for anodic hole generation (avalanche mechanism), thus the pore etching must stop [5].

Since the intersection of pores has been observed (Figure 4.5a), it is obvious that even at distances smaller than twice the space-charge region a sufficient number of holes is generated by avalanche breakdown at least to initiate the rate-limiting step of the dissolution reaction. As shown in Figure 4.6, when two pores meet in space the field strength around the pore tips is slightly reduced since the effective space-charge region width $W_{\text{eff.}}$ near the pore tip is enlarged. Decreasing d even further the band bending will be reduced and $W_{\text{eff.}}$ will also enlarge further but still the field strength can be high enough to allow for avalanche breakdown, i.e., the distance d between the pores is not the decisive parameter to describe the electric field strength necessary to generate holes in n-type semiconductors and therefore the intersection of pores is possible.

The subsequent reaction steps are now easy since the oxidation intermediates produced by this initial reaction step are usually considered to generate surface states with energy levels above the valence band edge [45]; they can thus be easily oxidized by injection of an electron into the conduction band by tunnelling or thermal activation as explained above.

Our experiments were carried out under conditions of strongly anisotropic etching along guidelines adopted by the "current-burst model" proposed for pore formation in Si [46]. The strong influence of the crystal anisotropy under those conditions probably also plays a major role in the crossing of pores and subsequent continuation of the growth along the same crystallographic directions.

4.3.3. Current-Line-Oriented Pores

Figure 4.7 presents the cross-section micrograph of a (100)-oriented n-InP sample anodized at the current density $j = 60$ mA/cm^2. Pores obtained in these conditions

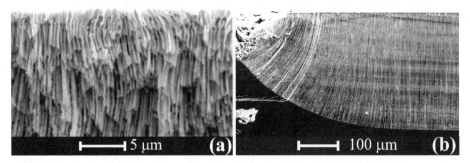

FIGURE 4.7. Current-line-oriented pores obtained at high anodic current densities: (a) far from the O-ring, (100)-oriented InP, $n = 3 \times 10^{17} \text{cm}^{-3}$ and (b) near the O-ring, (100)-oriented InP $n = 10^{18} \text{cm}^{-3}$.

obviously have no specific crystallographic orientation but simply follow some curved lines mainly oriented perpendicular to the surface. The system seems to minimize the ohmic loses by optimizing the way for the current [47]. We call such pores "current-line oriented".

As it was shown above, under nearly identical conditions except for a lower value of the anodic current density of more than one order of magnitude, a totally different morphology is obtained (Figure 4.5b). In this case, the pores prove to be strictly oriented along specific crystallographic directions.

Taking into account that the results described above are well reproducible, one can conclude that a gradual change of the current density from low to high values leads to a switch of the pore growth mechanism from being crystallographically oriented to current-line oriented. This conclusion is supported by the images presented in Figures 4.5b and 4.7.

In spite of the fact that the current-line-oriented pores grow preferentially perpendicular to the surface, they show a pronounced curvature in their direction of growth near the O-ring of the electrochemical cell (see, e.g., Figure 4.7b). Also, it is worth noting to mention that the instability of the direction of pore growth at high current densities is more pronounced when the anodic current through the sample is applied in pulses, this means that the current is a factor influencing the path of the growth for the current-line-oriented pores.

So far, current-line-oriented pores have been observed at high current densities only in n-type (100)-oriented InP and GaP, but not in GaAs. In GaAs at high current densities, either the so-called tetrahedron-like pores begin to grow, or the domains of crystallographically oriented pores are formed.

4.3.4. Tetrahedron-Like Pores in GaAs

Figure 4.8 shows a SEM image in cross section taken from a porous layer fabricated at a current density of 5 mA/cm^2. In this case, with the exception of the phase of pore nucleation, the monitored voltage (under galvanostatic control) between the RE and SE electrodes has a relatively constant value during the whole duration of the experiment. Being oriented along $\langle 111 \rangle$ directions, the pores possess a well-developed triangular prism-like shape (see the inset in Figure 4.8) with $\{112\}$ planes as facets [11].

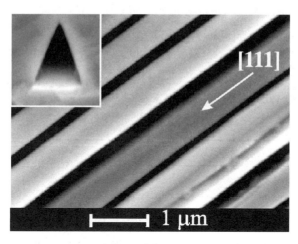

FIGURE 4.8. Smooth pores oriented along ⟨111⟩ directions obtained in (100)-oriented GaAs ($n = 10^{17} cm^{-3}$) at low current densities.

However, the time dependence of the RE–SE voltage changes significantly when the current is sufficiently increased. After an initial interval of relatively constant voltage at the beginning of the etching process, an increase in the voltage occurs. At the current density $j = 85$ mA/cm^2, for instance, a sharp voltage jump occurs after approximately 75 minutes of etching in 5% HCl. SEM investigations revealed that the voltage jump initiates a new phase of pore growth. Figure 4.9 shows SEM images in cross section taken from a sample anodized at a current density of 85 mA/cm^2. The pores shown were observed after the voltage jump and still form an angle of 54° with the normal to the surface. This means that they continue to grow along ⟨111⟩ crystallographic directions, as they did before the voltage jump. However, the pore walls are not smooth anymore as those presented in Figure 4.8, but form the afore-mentioned chains of interconnected tetrahedral voids which clearly indicate a self-induced diameter oscillation of individual pores. Three of the four facets of the tetrahedron can be easily observed in Figures 4.9a and b.

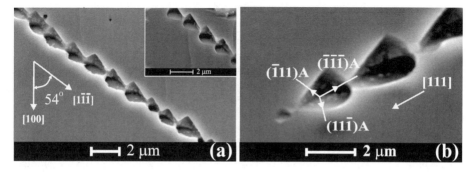

FIGURE 4.9. Tetrahedron-like pores oriented along ⟨111⟩ directions obtained in (100)-oriented GaAs at high current densities in HCl electrolytes ($n = 10^{17} cm^{-3}$).

It is worth noting to mention that a similar morphology has been observed earlier in Si [48], but instead of tetrahedron-like voids, an octahedron-like structure develops in Si because the "stopping planes" as described below are different from those in GaAs. The "current-burst" model explains the octahedron-like pores observed in Si by means of the so-called ageing of the pore walls. Ageing in this case describes the passivation at pore tips with time as a function of the current density. H passivation hinders the oxidation of Si and is greatly dependent on crystallographic planes. {111} planes are the most easy to passivate and thus the most stable against dissolution, therefore the octahedron-like pores expose them as "stopping planes".

The observation of self-induced oscillations of the pore diameter in quite different semiconductors such as Si and GaAs indicates that the general mechanism governing pore formation in these materials is not so different. Thus, we will try to use the "ageing" concept introduced for Si to explain the formation of the tetrahedron-like pores observed in GaAs taking into account that in GaAs it is Cl passivation rather than H passivation that impedes oxidation and thus the process of local dissolution [49]. Moreover, the "stopping planes" in this case are the so-called {111}A planes, i.e., {111}Ga-rich planes. Therefore, it is expected that the observed tetrahedron-chain pores in GaAs expose these planes.

Briefly, the "ageing" concept in Si comprises the following basic principles. Current flow is supposed to be inhomogeneous in space and time, i.e., there are "current bursts" followed by periods without charge transfer during which passivation may occur. Passivation of pore tips is thus more pronounced at lower current densities because there is more quiescent time between current bursts. In the case of III–V compounds, it can be assumed that each avalanche breakdown can be viewed as a current burst on a somewhat larger scale. As discussed above, the avalanche breakdown can also start and stop like a current burst and is greatly influenced by the surface passivation which can neutralize the surface defects and thus decreasing the probability for a new avalanche breakdown event to occur.

A schematic description of the process assumed to be responsible for the formation of tetrahedron-like pores is presented in Figure 4.10.

Consider the situation when the current density at a pore tip (j_{tip}) has reached a critical low value j_{min} where nearly the whole surface of the pore tip (A_{tip}) is passivated ($t < t_1$ in Figure 4.10). The mechanism how the system reaches the critical current density will be explained later. Having in mind that the experiments were performed in a galvanostatic regime the current must flow through the sample all the time and in all conditions. Thus, it can happen that in order to maintain a constant high current through the whole sample, the system will be forced to "abandon" a significant part of the strongly passivated surface of the tip and to concentrate the current flow only through a small area at the tip of the pore in order to increase the current density which in turn decreases significantly the passivation ($t = t_1$ in Figure 4.10). Therefore in the diagram in Figure 4.10, at $t = t_1$ sharp steps for S_{tip} (down) and j_{tip} (up) occur. The reason why the system chooses the small end of the tip is that here the electric field strength is higher due to the curvature of the pore tip and it is easier to break the passivation. Immediately after t_1, a small spherical cavity at the tip of the pore is formed (see Figure 4.9b). The cavity does not expose any crystallographic planes, no preferential passivation is present and the reaction in this stage can be considered to be kinetically controlled. This means that due to dissolution the surface of the sphere will increase

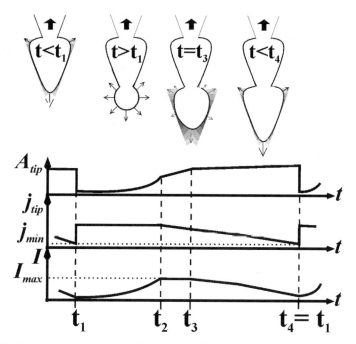

FIGURE 4.10. Schematic representation of the process explaining the tetrahedron-like pore formation according to the "ageing" concept.

with time in a quadratic manner (sphere), while the current density at the tip remains constant ($t_1 < t < t_2$). Consequently, during this period of time the value of the current $I = j_{tip} S_{tip}$ will increase as the surface does. Nevertheless, taking into account that the tetrahedron-like pores begin to grow far underneath the surface, the current I will increase only up to a certain maximum value (I_{max}) which in fact is defined (limited) by the diffusion of the species (reducing, oxide dissolving and reaction products) to and from the pore tips. Because I_{max} is the maximum current value which at this moment the system can transport through one pore, it will try to keep $I = I_{max}$ constant for a while ($t_2 < t < t_3$). Since the surface of the sphere will continue to grow (dissolution takes place), the current density starts to decrease and consequently according to the "current-burst" model the surface passivation will increase. $\{111\}$A planes will be passivated more easily in this case. Thus, after the diffusion limited current is reached and the current density at the pore tip begins to decrease again, the spherical cavity will begin to expose these planes ($t > t_2$) transforming itself into a tetrahedron-like cavity. The current density j_{tip} will continue to decrease until it reaches again the critical value and consequently a new step in A_{tip} and j_{tip} will occur ($t = t_4$), i.e., a new tetrahedron begins to grow.

As can be observed from Figure 4.10, the current at a pore tip I varies with time, following the phase of the growth of the tetrahedrons. Note that in the "current-burst" model developed for Si, one of the general assumptions is that the current at the pore tip also oscillates but due to random phases of the oscillations at different pore tips a constant macroscopic current across the sample can be obtained.

Now the answer to the question why tetrahedron-like pores are observed only after a definite interval of time from the beginning of the experiment is relatively simple: At the beginning of the experiment the current density at pore tips is much higher than the critical value j_{min}, therefore the system needs some time before reaching it by successive branching of the original pores.

The second question is why the system can reach this critical value at pore tips only at high externally applied currents? It has been observed earlier that the pores initially nucleated at the surface of the sample are branching during the anodization process [11]. If a constant external current is applied to the sample, the current density at pore tips depends on the number of pores in the substrate. Thus, due to branching, the number of pore tips in the substrate increases with time, which means that the current density at the pore tips is not constant but decreases with time. Eventually, the decrease of j_{tip} with time allows the system to reach the critical j_{min} current density at the pore tips which leads to the formation of linked tetrahedrons as explained above. At low values of the external current the branching effect is practically absent or strongly reduced, therefore the number of pore tips is constant with time and so is the current density at the tips, i.e., it cannot decrease to the critical value where the tetrahedron pores begin to grow. The second reason is that the diffusion limited current I_{max} at low external current densities cannot be achieved because the amount of dissolved material is small as compared with the case of high external current densities.

In order to test the above assumptions, the influence of the electrolyte concentration and current density on the time interval from the beginning of the experiment until tetrahedron-like pores begin to grow was studied. Figure 4.11 shows the time dependence of the voltage at different current densities for two electrolyte concentrations, 5% HCl (Figure 4.11a) and 10% HCl. (Figure 4.11b). The time needed to reach the tetrahedron regime of pore growth shortens both by increasing the external current density and the electrolyte concentration.

These observations can be explained by the higher number of Cl^- ions in concentrated electrolytes. Thus, the number of passivated bonds per unit of time is larger in comparison with more diluted electrolytes. Comparable passivation states (equal number of passivated bonds on surface unit) for two different concentrations of the electrolyte

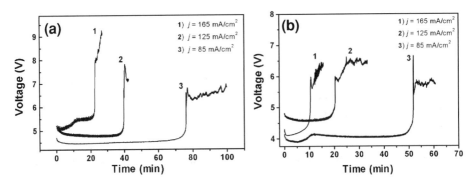

FIGURE 4.11. Time dependence of the voltage across the sample during the anodization process at different external current densities: (a) HCl 5% and (b) HCl 10%.

$c_1 < c_2$, are obtained for two different current densities $j_1 < j_2$ that must flow through pore tips. Consequently, the critical value j_{min} will be higher in more concentrated solutions ($j_{min1} < j_{min2}$). As was explained above, the actual current density at pore tips decreases with time (due to branching) from the initial value j_{init} at the surface of the sample until j_{min} is reached and the system enters a new state (voltage jump). So, the time needed to reach j_{min2} is shorter than that required to reach j_{min1} because $j_{init} - j_{min2} < j_{init} - j_{min1}$. This is in good agreement with the results presented in Figures 4.11a and b.

The decrease of the time necessary for the system to reach the tetrahedron-like pore regime when increasing the externally applied current density can be explained taking into account the fact that branching of pores occurs more frequently at high externally applied currents and is practically absent at low currents. Thus, j_{tip} decreases faster at high than at low values of the externally applied current. Consequently, j_{min} is reached after a shorter time.

The observed architecture of pores offers new insights into the mechanism of pore formation in GaAs, and also represents an interesting variety of possible pore structures for applications. The pores obtained at high current densities after the voltage jump look like asymmetrically modulated microchannels and, according to the new concept of drift ratchet [50], could be suitable for designing micropumps to separate micrometre-size particles dissolved in liquids.

4.4. SELF-ORGANIZED PROCESSES DURING PORE FORMATION IN III–V COMPOUNDS

4.4.1. What is Self-Organization?

The essence of self-organization is that the system structure appears without explicit pressure or constraints from outside the system. Or otherwise stated, the constrains on form are internal to the system and result from the interaction between the components, whilst being independent of the physical nature of those components. The organization can evolve in space and time, can maintain a stable form, or can show transient phenomena.

Many observations made during pore formation processes in III–V compounds, proved to have characteristics of self-organization. One of the examples is the already discussed tetrahedron-like pore in GaAs which begin to grow without any special constrains from exterior. Here, it is important to note that the "external current" mentioned in the text is considered an inherent part of the system. As will be shown later on, there are more examples confirming the tendency of self-organization or self-ordering during pore formation.

4.4.2. Voltage Oscillations: An Emergent Property at High Density Pore Growth

In the etching experiments carried out under galvanostatic conditions an interesting behaviour of the voltage measured between the n-InP sample ($n_1 = 1.5 \times 10^{16}$ cm^{-3}) and a Pt electrode in the electrolyte (RES and REE) was found. At the beginning of the

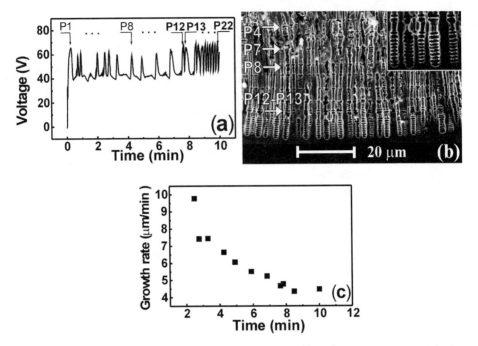

FIGURE 4.12. Data taken from an InP sample with $n_1 = 1.5 \times 10^{16}$ cm^{-3} anodized at a current density $j = 100$ mA/cm^2. (a) Voltage oscillations; (b) cross-sectional SEM of the sample. The inset is the magnification of the nodes at the bottom of the porous layer and (c) the rate of pore growth during the anodization of InP with $n = 1.5 \times 10^{16}$ cm^{-3}.

anodization process, the voltage increased monotonously from 0 to about 60 V. After reaching the maximum value, the voltage started to oscillate as illustrated in Figure 4.12a. The cross-sectional SEM image of a porous sample obtained under these etching conditions is shown in Figure 4.12b. The diameter of the pores is strongly modulated, the observed local increases in the pore diameter will be called pore "nodes". It is apparent that nodes of different pores are correlated as a function of depth by the "trajectories" connecting nodes obvious to the eyes of the observers. A comparative analysis of Figures 4.12a and b shows that each line of nodes is directly coupled to one voltage maximum. Consequently, there is a direct relation between the peaks visible on the voltage/time diagram (Figure 4.12a) and the formation of pore nodes. The voltage maxima and the corresponding pore nodes are marked in Figures 4.12a and b as "P1",..., "P22" (note that Figure 4.12b shows only the nodes from "P4" to "P22").

The correlation between the voltage maxima and pore nodes offers the possibility to calculate the rate of the pore growth. From Figure 4.12a one can calculate the time interval between two voltage maxima, while from Figure 4.12b it is possible to estimate the growth of the pores during the period involved.

Figure 4.12c shows that the rate of pore growth decreases from nearly 10 μm/min at the beginning to approximately 5 μm/min at the end of the etching experiment. The retardation of pore growth, in spite of the fact that the electrolyte was continuously pumped through the cell, may be caused by the increasingly difficult transport of chemical

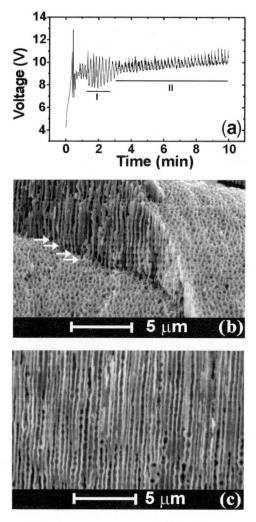

FIGURE 4.13. Data taken from an InP sample with $n_2 = 3 \times 10^{17}\text{cm}^{-3}$ anodized at a current density $j = 100$ mA/cm^2. (a) Voltage oscillations; (b) SEM image and (c) cross-sectional SEM taken from an InP sample with $n_2 = 3 \times 10^{17}\text{cm}^{-3}$ anodized under conditions of small-amplitude voltage oscillations.

species from and to the pore tips by diffusion. At the beginning of the experiment (etching in the vicinity of the initial surface) more oxide dissolving species were available for the dissolution while in depth the number of oxide dissolving species provided by the process of diffusion decreases and, as a consequence, the dissolution rate goes down.

When anodizing n-InP samples with the higher doping level $n_2 = 3 \times 10^{17}$ cm^{-3}, voltage oscillations were also observed, as shown in Figure 4.13a. The amplitude of the oscillations varies considerably during the time of anodization. High amplitude voltage oscillations (region I in Figure 4.13a) can be traced to a synchronous modulation of the pore diameters. In this case, the SEM images in cross section show weak horizontal trajectories illustrated in Figure 4.13b: Both horizontal trajectories (marked by arrows) in cross section and the morphology of the porous layer as a function of depth can be seen. As expected, the layer morphology does not depend much on the depth of the pores. The correlation between pores, however, seems to be lost at larger depths.

A detailed SEM image in cross section taken from an area corresponding to region II with smaller voltage amplitudes (Figure 4.13a) is presented in Figure 4.13c. Only a small part of the pores shows synchronous modulation of their diameters under the etching conditions involved and trajectories are not apparent. This does not necessarily indicate a total loss of synchronization between pores, however, since the correlation may either be too weak to be directly visible or it may only occur in regions not contained in the picture.

The observed self-induced voltage oscillation can be considered as an emergent property of strongly interacting pores. This means that the oscillations will immediately disappear if the pores stop to interact.

4.4.3. A Model Explaining the Voltage Oscillations

In contrast to current oscillations which may occur locally under constant voltage conditions at pore tips (or, more generally, in arbitrarily large domains) but add up to a constant external current if the phases are distributed randomly, local voltage oscillations are not possible—the voltage along any path between the electrodes must be the same. Voltage oscillations under constant external current conditions may be understood if we assume that the current at the pore tip generally oscillates while the diameter stays nearly constant, i.e., the current density oscillates too. Both assumptions which we adopt in general without specifying the current oscillation mechanism at this point (which may be different from the oxide-based Si case) [51,52] are general properties of the current-burst model for Si. In equivalent circuit terms each pore may then be described by an oscillating resistor $R(t)$ with the average value $\langle R \rangle$ (Figure 4.14). The total current is given by switching all resistors in parallel to the voltage/current source. As long as the phases of the oscillating resistors are uncorrelated, i.e., random, the total current will have some constant average value given by $\langle I \rangle = U/\langle R \rangle$ (where U is the voltage).

We consider each pore as a resistor defined as $R = \rho L/A$, where ρ is actually the resistivity (resistance per unit length per unit area) of the oxide layer formed at the tip and A is the surface of one pore tip. Taking into account that the thickness of the oxide

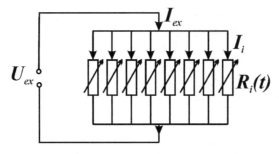

FIGURE 4.14. A schematic representation of pores as oscillating resistors.

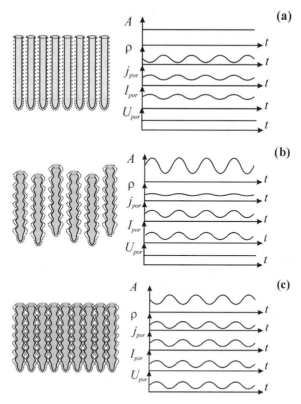

FIGURE 4.15. A schematic representation of pore interaction explaining the morphologies observed in III–V compounds: (a) ρ oscillates and A is constant; (b) ρ is constant and A oscillates; and (c) ρ and A oscillate simultaneously.

layer (L) equals only a few angstroms we will consider it constant. Thus, oscillation of R can be attributed only to the variation in time of ρ and A. From this consideration three cases can be significant: (a) ρ oscillates and A is constant; (b) ρ is constant and A oscillates and (c) ρ and A oscillate simultaneously. Actually the system is varying these parameters in time aiming at minimizing the resistance of the entire sample. All these three cases are presented schematically in Figure 4.15.

In Figure 4.15a, the case when A is constant but ρ oscillates is presented. Due to the fact that ρ and respectively R oscillates, I_{por} and j_{por} will also oscillate. It is to be noted that in a galvanostatic regime (I_{tot} = const.), the oscillation of I_{por} will be possible only if I_{por} of each pore in the system oscillates with a random phase shift relative to its neighbours. Thus, the summation of all randomly oscillating currents in the system will result in a constant externally flowing current I_{tot} = const. The condition of randomness can be fulfilled only when the pores do not interact with each other, i.e., the distances between them is larger than the double thickness of the space-charge layer and they do not have the possibility to synchronize themselves. Examples of such pores are the ones presented in Figure 4.5 where the diameter of the pores is constant and the distance between them is larger than the double value of the space-charge layer.

The case (b) is illustrated in Figure 4.15b where oscillation of A is dominant over the oscillations of ρ which oscillates only slowly. Also here, the current and current density at the tip of the pores oscillate as R does and the constant external current imposed by galvanostatic conditions is reached due to the randomness in current oscillations of each pore. In this case, too, the pores should not feel each other and the distance between them should be rather large. Such kinds of pores have been observed in GaAs (see tetrahedron pores).

In the third case, the pores are so close to each other that they begin to interact via the space-charge region. Interaction will limit the amplitude of A oscillations. The system no longer has the freedom—as before—to minimize the resistance of the sample by varying A only, therefore ρ must start to oscillate, too (see Figure 4.15c). The interaction between pores now may influence the whole system, and a correlation (synchronization) between a certain number of pores can be achieved. Synchronization of pores in some parts of the sample (domains) can only be achieved by correlating the phases of the oscillating resistors while $R(t)$ remains the same (in this simple approximation). The total current through a domain now will no longer average out to a constant value, but will also oscillate and the constant current condition enforced by the external current source now can only be maintained if the voltage oscillates so that $I_{domain} = const. = NU(t)/R(t)$, where N denominates the number of pores in the domain.

This simple model explains all observations. Of course, in a better approximation one would have to describe a pore by a more complex equivalent circuit containing capacitors C. The displacement currents $I_{cap} = C \cdot (dU/dt)$ must also be compensated by the voltage adjustments, causing some degree of feedback in the system. Nevertheless, the ultimate causes of the voltage oscillation are intrinsic current oscillation together with some phase coupling or correlation between pores.

This consideration, if turned around, gives a clue to the interaction mechanism between neighbouring pores. If, by random fluctuation of pore diameters, pores come close enough to experience some kind of influence on their states of dissolution, a feedback mechanism may start that leads to phase coupling of the pore states and by percolation to the formation of a synchronized domain.

This domain, however, may cover only a part of the specimen surface, i.e., regions with uncorrelated pores may also be found. Moreover, since percolation does not have to take place at every cycle of the oscillation—especially if the general conditions are just around the percolation point of the system—somewhat irregular voltage oscillations as shown in Figure 4.13a are possible. The frequency of the voltage oscillations in this model would be determined by the frequency of the current oscillations inherent to the current-burst model.

4.4.4. Quasi-Uniform Self-Arranged Pores in InP

As was shown above, the observed external voltage oscillations were coupled to pore diameter oscillations. This means that by controlling one of these parameters, the second one can be controlled, too. Thus, in order to achieve porous arrays with smooth pores it is necessary to anodize in the potentiostatic regime rather than in a galvanostatic one.

Figure 4.16a presents a random cleavage of a porous layer produced by anodic etching of an n-InP substrate at constant voltage $U = 10$ V for 1 minute. After removal

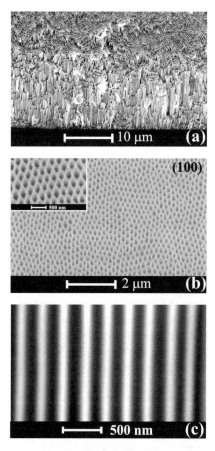

FIGURE 4.16. SEM micrographs of an InP sample exhibiting self-arranged CLO pores under potentiostatic conditions: (a) random cleavage overview and (b) top view after the NL has been removed.

of the nucleation layer (NL, not seen in Figure 4.16a) by nonselective wet etching or by mechanical polishing, the distribution of porosity was found to be highly uniform both along the top surface and in the depth of the porous layer. The NL with the thickness of 1.5 µm and a degree of porosity less than 10% exhibited the so-called CO pores aligned along $\langle 111 \rangle$ directions. The thickness of the NL was found to depend upon the anodization conditions, namely increasing the concentration of the electrolyte led to an increase in the NL thickness.

After the formation of the NL, further anodization under the same conditions favours the propagation of pores perpendicularly to the initial surface (Figure 4.16a), i.e., along the current lines [12]. From a high magnification image it can be observed that now indeed the diameter oscillations are absent and the walls of the pores are very smooth (Figure 4.16c). We believe that interaction between pores by means of the space-charge region as explained above, as well as the high crystalline quality of our samples, and the computer-controlled accuracy of etching conditions are the primary factors leading to the observed uniformity in pore distribution. Moreover, SEM top views of the sample taken after the removal of the NL show the spatial distribution of pores to be ordered (Figure 4.16b). In fact, according to the inset presented in Figure 4.16b, the porous layer represents a 2D triangular lattice of pores embedded in InP matrix. The transverse sizes of pores and InP wall thickness equal 200 and 120 nm, respectively, leading to a degree of porosity of about 55%. An autocorrelation analysis of the pore positions shows an ordering of up to the sixth neighbour. A corresponding analysis of the radial distribution of the pore position shows a prevalent 60° angle between the neighbours of one pore. Combining both data indicates a respectable long-range ordering of pores in a closed-packed array.

4.5. POSSIBLE APPLICATIONS OF III–V POROUS STRUCTURES

4.5.1. Bragg-Like Structures on n-InP

Two different types of pores have been observed until now in n-InP: crystallographically oriented (low currents) and current-line-oriented pores (high currents). The two types of morphologies have quite different degrees of porosity and thus different refractive indices. In consequence, a periodic change from one type of morphology to the other would result in the formation of Bragg-like structures on InP substrates.

In order to change the degree of porosity as a function of depth, the process of anodic etching was periodically switched on and off, e.g., from $I = 600$ mA/cm^2 (duration $t_1 = 0.1$ minutes) to $I = 0$ mA/cm^2 (duration $t_2 = 0.5$ minutes) and vice versa. It is obvious that the etching itself takes place only during the first 0.1 minutes of the cycle, while in the last 0.5 minutes no dissolution occurs (no current flows).

The general finding is that a process interruption for 0.5 minutes is enough for the system "to loose its memory". When the current is switched on again, a new nucleation phase is required, i.e., a new NL with CO pores emerges before the formation of CLO pores. Note that this fact is not understandable in any static pore formation model where the equilibrium structure, once reached, should be maintained under all conditions. This observation, however, fits exactly in the general scheme of the current-burst model because in the current-free phase the pore tips become passivated and the self-ordering process of current bursts has to be started again.

Applying this procedure, a structure was obtained with a spatially modulated degree of porosity consisting, e.g., of 32 alternating layers of high and low porosity (Figure 4.17). The difference in degree of porosity is also confirmed by cathodoluminescence (CL) measurements (Figure 4.18).

SEM and CL images have been taken from the same cleavage region of a Bragg-like structure. As can be observed from Figure 4.18b, the CL proves to be spatially modulated; the CL intensity from NL layers being higher than that from CLO layers. Taking into account that the CL intensity from bulk InP is higher than that from a porous layer, it can be concluded that a porous layer with a higher CL has a lower porosity (more similar to the bulk) than the one with a lower CL. Different porosities mean also different refractive indices. Thus, porous layers with CO pores have a higher refractive index than those formed by CLO pores.

Although neighbouring layers exhibit quite different morphologies and degrees of porosity, the interfaces prove to be rather sharp, see the inset in Figure 4.18a. Note that

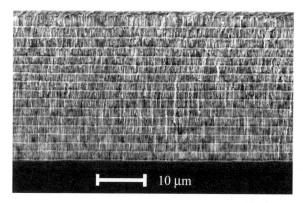

FIGURE 4.17. Cross-sectional SEM micrograph of a porous n-InP Bragg-like structure with spatially modulated degree of porosity.

the CO pores in the NL possess a triangular shape with transverse size less than 100
nm. As to the CLO pores, they may have nearly quadratic or circular shapes with the
transverse sizes varying from 50 to 700 nm as a function of the electrolyte concentration,
current density and substrate doping.

In summary, the obtained results show the possibility to design and fabricate Bragg-
like structures on n-InP substrates.

4.5.2. Towards Uniform 3D Structures in GaAs

As mentioned above, holes necessary for pore formation in III–V compounds are
generated by a breakdown mechanism, and the dissolution (nucleation) usually starts at
defects. In order to obtain a uniform porous structure suitable as photonic material, it

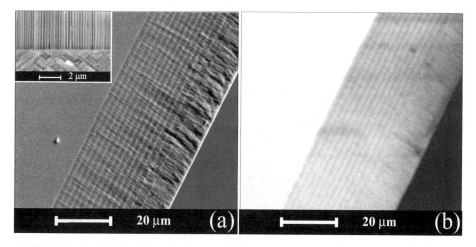

FIGURE 4.18. (a) SEM and (b) panchromatic CL images in cross section taken from a porous InP structure
with spatially modulated degree of porosity. The inset shows the interface between two neighbouring porous
layers.

should therefore be possible to use a predefined uniform network of defects (etch pits) as starting points for the pores.

If the pores can be nucleated in this way, further they must be forced to grow stable along definite directions which is possible only if optimized anodization conditions are found. The stable directions of growth are normally defined by the etching anisotropy, and in III–Vth these crystallographic directions are ⟨111⟩.

There are several open questions in this approach, and very little is known about them up to now. Some first answers we will give in the following.

First, we note that pore growth along ⟨111⟩ directions may lead to the formation of a 3D dielectric structure. Note that ⟨111⟩-oriented pores intersect each other without changing their direction of growth or shape (Figure 4.5), which is the essential for producing 3D periodic dielectric structures by intersecting the quadruples of ⟨111⟩ pores starting from a common nucleus in analogy to the ⟨113⟩ pores in Si forming the "Kielovite" structure [53].

However, experimentally it is not simple to fabricate such 3D structures. First, because the four directions are not equivalent from the chemical/electrochemical point of view. Two of them, the so-called ⟨111⟩B directions, have a lower dissolution rate than the other two ⟨111⟩A directions. There are two possible solutions to these problems: (i) special etching conditions should be chosen in order to reach the same dissolution rate in all four directions, or (ii) the dissolution along "slow-dissolving" directions could be completely inhibited.

Concerning the first approach, different methods (etching conditions) have been used in order to achieve equal dissolution rates for all four ⟨111⟩ directions. Among them was an attempt to manipulate externally the etching process by using periodic pulses of high currents. The idea is to increase the probability of the branching of pores and thus to achieve uniform pore growth in all four directions. First experimental results are presented in Figure 4.19a.

Although the pore structure in Figure 4.19a shows a nearly periodic arrangement of the ⟨111⟩A pores reflecting the periodicity of the externally applied current pulses, the ⟨111⟩B pores (perpendicular to the plane of the picture and visible as small holes) grow quite inhomogeneously with a strongly reduced density.

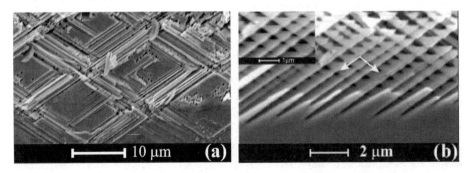

FIGURE 4.19. (100)-oriented n-GaAs, cross section: (a) a random 3D structure obtained by using high current pulses and (b) a quasi-uniform 3D structure obtained by using high constant currents.

In the second approach, the number of pores growing in ⟨111⟩B directions at moderate current densities decreases as the density of nucleated ⟨111⟩A pores increases. This means that when the ⟨111⟩A pores are close to each other, the nucleation of ⟨111⟩B pores is strongly inhibited. An example of high density randomly nucleated pores is shown in Figure 4.19b, where only two directions of pore growth are prevalent. Note that the triangles visible are not due to (inclined) ⟨111⟩B pores, but result from the intersection of ⟨111⟩A oriented pores.

It thus appears to be possible to find condition assuring stable growth of pores only in ⟨111⟩A directions once they are nucleated. It remains to be seen, if nucleation can be induced by providing lithographically defined defects. Electron-beam lithography has been used to pattern (100)-oriented GaAs wafers with the arrays of windows in the resist with diameters of 300 nm. Patterns with triangular and square lattices have been used. The lattice constant varied from 300 nm to 2 μm.

A top and cross-sectional view of the first attempts to anodize prepatterned samples is presented in Figure 4.20.

One can see (Figures 4.20a–c) that the pores indeed nucleate in the windows defined by the electron-beam lithography. In each window, two small black spots can be observed separated by a white band. The black spots are the nucleated ⟨111⟩A pores. A cross section of two pores nucleated from one window is presented in Figure 4.20d. In spite of the fact that nucleation in predefined windows seems to work, some other problems

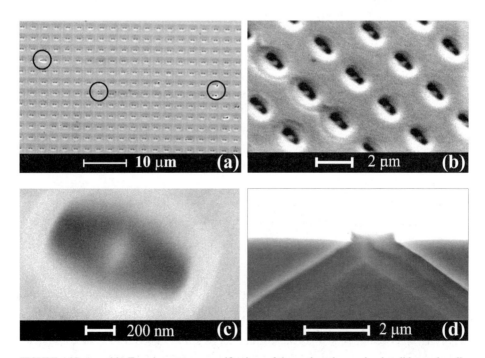

FIGURE 4.20. (a and b) Top view at two magnifications of the nucleated pores in photolithographycally defined windows and (c) high magnification of the nucleated pores. The two ⟨111⟩ directions can be observed; (d) cross section view of the nucleated pores in one window.

can appear during the etching process. For example, in addition to the pores nucleated in predefined windows some random pores nucleate too. Such randomly nucleated pores are marked in Figure 4.20a by black circles. It is likely that random pores nucleate at defects with a lower breakdown voltage, leading to faster nucleation, faster growth, high current densities and successive branching, diminishing the necessity to nucleate or sustain pores in the predefined windows in order to carry the external current. Therefore, even the pores which have started to nucleate in predefined windows can stop growing in favour of the randomly nucleated pores.

4.6. CONCLUSION

Pore etching in III–V compounds is dominated by a series of self-organized processes:

- Interactions in space can explain the following effects:
  - Formation of a nearly closed-packed 2D array of pores in InP with an ordering up to the seventh neighbour.
  - Self-induced simultaneous diameter oscillations of all pores in InP allowing for a quite simple and stable structuring of a quasi 2D pore array.
  - Intersection of crystallographically oriented pores allowing for the formation of a real 3D pore crystal in GaAs.
- Interactions in time can explain the following effects:
  - Formation of chains of tetrahedron-like pores in GaAs.
  - Octahedrons in InP which are not as well defined as tetrahedrons in GaAs but still diameter modulations for individual pores are found which by increasing the current density start to synchronize. First, only small domains of synchronized pore diameter are found until at high current densities all pores grow phase coupled.

Due to this high complexity of nonlinear interactions many different pore morphologies can be found in one system, just by changing one parameter, e.g., the current density:

- Crystallographically oriented pores change to current-line-oriented pores.
- Pores with smooth walls change to tetrahedron-like pores.

Critical points exist for the switch from one growing phase to another, indicated e.g., by a voltage jump.

Many of the self-organized processes observed in III–V compounds can be explained by the current-burst model developed first for Si. Due to this model, current always flows in local bursts consuming a defined amount of charge. Changing the current density therefore mainly changes the density of current bursts and the time for passivation between two subsequent current bursts at the same position of the sample. This model in principle holds for pore formation in all semiconductors. The differences in the pore morphologies for different III–V compounds depend on the stability of the generated oxide and the passivation properties of the semiconductor electrolyte interface which, e.g., allows us to understand the crystallographic dependence of pore formation as a function of the electrolyte. Due to the "chaotic" behaviour of the pore formation moderate differences in the reaction kinetics may lead to extreme differences in the resulting pore morphologies.

Since the electrochemically etched pore in III–V compounds show a great potential for applications (e.g., highly ordered 2D and 3D pore arrays for photonic application), a fundamental understanding of the underlying basic electrochemistry as well as the interactions between pores is very important. The results presented in this chapter show that pore formation is an interplay of electrochemistry, semiconductor and statistical physics. Controlled pore formation is therefore a big challenge and will need several years of research although promising results already exist.

REFERENCES

[1] W.P. Gomes and H.H. Goossens, Electrochemistry of III–V compound semiconductors: dissolution kinetics and etching, in *Advances in Electrochemical Science and Engineering*, Vol. 3, Ed. H. Gerischer and C.W. Tobias, VCH, Weinheim, 1994.

[2] P.H.L. Notten, J.E.A.M. van den Meerakker and J.J. Kelly, *Etching of III–V Semiconductors: An Electrochemical Approach*, Elsevier, Oxford, 1991.

[3] P. Schmuki, J. Fraser, C.M. Vitus, M.J. Graham and H.S. Isaacs, Initiation and formation of porous GaAs, J. Electrochem. Soc. **143**(10), 3316–3322 (1996).

[4] P. Schmuki, D.J. Lockwood, H.J. Labbe and J.W. Fraser, Visible photoluminescence of porous GaAs, Appl. Phys. Lett. **69**, 1620–1622 (1996).

[5] P.C. Searson, J.M. Macaulay and F.M. Ross, Pore morphology and the mechanism of pore formation in n-type silicon, J. Appl. Phys. **72**, 253–258 (1992).

[6] T. Takizawa, Sh. Arai, M. Nakahara, Fabrication of vertical and uniform-size porous InP structure by electrochemical anodization, Japan. J. Appl. Phys. **54**(2), L643–L645 (1994).

[7] B.H. Erne, D. Vanmaekelbergh and J.J. Kelly, Morphology and strongly enhanced photoresponse of GaP electrodes made porous by anodic etching, J. Electrochem. Soc. **143**(1), 305–314 (1996).

[8] I.M. Tiginyanu, G. Irmer, J. Monecke, A. Vogt and H.L. Hartnagel, Porosity-induced modification of the phonon spectrum of n-GaAs, Semicond. Sci. Technol. **12**, 491–493 (1997).

[9] I.M. Tiginyanu, C. Schwab, J.-J. Grob, B. Prevot, H.L. Hartnagel, A. Vogt, G. Irmer and J. Monecke, Ion implantation as a tool for controlling the morphology of porous gallium phosphide, Appl. Phys. Lett. **71**(26), 3829–3831 (1997).

[10] S. Langa, J. Carstensen, M. Christophersen, H. Föll and I.M. Tiginyanu, Observation of crossing pores in anodically etched n-GaAs, Appl. Phys. Lett. **78**(8), 1074–1076, (2001).

[11] F.M. Ross, G. Oskam, P.C. Searson, J.M. Macaulay and J.A. Liddle, A comparison of pore formation in Si and GaAs, Phil. Mag. A **75**, 525–539 (1997).

[12] S. Langa, I.M. Tiginyanu, J. Carstensen, M. Christophersen and H. Föll, Formation of porous layers with different morphologies during anodic etching of n-InP, Electrochem. Solid-State Lett. **3**(11), 514–516 (2000).

[13] M.A. Stevens-Kalceff, I.M. Tiginyanu, S. Langa and H. Föll, Correlation between morphology and cathodoluminescence in porous GaP, Appl. Phys. **89**(5), 2560–2565 (2001).

[14] S. Ottow, V. Lehmann and H. Föll, Development of three-dimensional microstructure processing using macroporous n-type silicon, Appl. Phys. A **63**, 153–159 (1996).

[15] A.I. Belogorokhov, V.A. Karavanskii, A.N. Obraztsov and V.Yu. Timoshenco, Intense photoluminescence in porous gallium phosphide, JETP Lett. **60**(4), 274–279 (1994).

[16] A. Anedda, A. Serpi, V.A. Karavanskii, I.M. Tiginyanu and V.M. Ichizli, Time-resolved blue and ultraviolet photoluminescence in porous GaP, Appl. Phys. Lett. **67**(22), 3316–3318 (1995).

[17] K. Kuriyama, K. Ushiyama, K. Ohbora, Y. Miyamoto and S. Takeda, Characterization of porous GaP by photoacoustic spectroscopy: the relation between band-gap widening and visible photoluminescence, Phys. Rev. B **58**(3), 1103–1105 (1998).

[18] F. Iranzo Marin, M.A. Hamstra and D. Vanmaekelbergh, Greatly enhanced sub-bandgap photocurrent in porous GaP photoanodes, J. Electrochem. Soc. **3**, 1137–1142 (1996).

[19] E. Kukino, M. Amiotti, T. Takizawa and S. Arai, Anisotropic refractive index of porous InP fabricated by anodization of (111)A surface, Japan. J. Appl. Phys. **34**(1), 177–178 (1995).

[20] I.M. Tiginyanu, G. Irmer, J. Monecke and H.L. Hartnagel, Micro-Raman scattering study of surface-related phonon modes in porous GaP, Phys. Rev. B **55**(11), 6739–6742 (1997).

[21] I.M. Tiginyanu, V.V. Ursaki, Y.S. Raptis, V. Stergiou, E. Anastassakis, H.L. Hartnagel, A. Vogt, B. Prevot and C. Schwab, Raman modes in porous GaP under hydrostatic pressure, Phys. Status Solidi b **211**, 281–286 (1999).

[22] I.M. Tiginyanu, I.V. Kravetsky, G. Marowsky and H.L. Hartnagel, Efficient optical second harmonic generation in porous membranes of GaP, Phys. Status Solidi a **175**(2), R5–R6 (1999).

[23] I.M. Tiginyanu, G. Irmer, J. Monecke, H.L. Hartnagel, A. Vogt, C. Schwab and J.-J. Grob, Porosity-induced optical phonon engineering in III–V compounds, Mater. Res. Soc. Symp. Proc. **536**, 99–104 (1999).

[24] S. Li and J.B. Khurgin, Feasibility of phonon-assisted electronic devices, J. Appl. Phys. **74**(4), 2562–2564 (1993).

[25] W. Plieth and S. Witzenstein, Semiconductor micromachining: fundamental electrochemistry and physics, in *Semiconductor Micromachining, Volume 1: Fundamental Electrochemistry and Physics*, Vol. 1, Ed. S.A. Campbell and H.J. Lewerenz, John Wiley & Sons, 1998.

[26] S. Roy Morrison, *Electrochemistry at Semiconductor and Oxidized Metal Electrodes*, Plenum Press, New York, 1980.

[27] R.M. Osgood, A. Sanchez-Rubio, D.J. Ehrlich and V. Daneu, Localized laser etching of compound semiconductors in aqueous solution, Appl. Phys. Lett. **40**, 391–393 (1982).

[28] D.V. Podlesnik, H.H. Gilgen, R.M. Osgood, A. Sanchez and V. Daneu, *Laser Diagnostics and Photochemical Processing for Semiconductors*, Ed. R.M. Osgood, S.R.J. Brueck and H.R. Schlossberg, North-Holland Press, New York, 1983.

[29] H. Gerischer and I. Wallem-Mattes, Z. Phys. Chem. NF **64**, 187 (1969).

[30] P.A. Kohl, Photoelectrochemical etching of semiconductors, J. Res. Dev. **42**(5), 629–638 (1998).

[31] H. Föll, M. Christophersen, J. Carstensen and G. Hasse, Formation and application of porous silicon, Mater. Sci. Eng. R **39**(4), 93 (2002).

[32] C. Levy-Clement, A. Lajousi and M. Tomkievicz, Morphology of porous n-type silicon obtained by electrochemical etching, J. Electrochem. Soc. **141**, 958–967 (1994).

[33] M.M. Carrabba, N.M. Nguyen and R.D. Rauh, Effects on doping and orientation on photoelectrochemically etched features in n-GaAs, J. Electrochem. Soc. **134**(7), 1855–1859 (1987).

[34] H. Föll, Properties of silicon–electrolyte junctions and their application to silicon characterization, Appl. Phys. A **53**, 8–18 (1991).

[35] H. Gerischer, P. Allongue and V. Costa Kieling, The mechanism of the anodic oxidation of silicon in acidic fluoride solutions revisited, Ber. Bunsenges. Phys. Chem. **97**, 753–757 (1993).

[36] P. Schmuki, L.E. Erikson, D.J. Lockwood, B.F. Mason, J.W. Fraser, G. Champion and H.J. Labbe, Predefined initiation of porous GaAs using focused ion beam surface sensitization, J. Electrochem. Soc. **146**, 735–740 (1999).

[37] G. Oskam, A. Natarajan, P.C. Searson, F.M. Ross, J.M. Macaulay and J.A. Liddle, *State of the Art Program on Compound Semiconductors XXI*, Electrochemical Society, Pennington, New Jersey, 1995.

[38] T. Takebe, T. Yamamoto, M. Fujii and K. Kobayashi, Fundamental selective etching characteristics of $HF + H_2O_2 + H_2O$ mixtures for GaAs, J. Electrochem. Soc. **140**(4), 1169–1180 (1993).

[39] D. Tromans, G.G. Liu, F. Weinberg, The pitting corrosion of p-Type GaAs single crystals, Corrosion Sci. **35**, 117–125 (1993).

[40] J.C. Tranchart, L. Hollan and R. Memming, J. Electrochem. Soc. **125**(7), 1185 (1978).

[41] E. Yablonovitch, T.J. Gmitter and K.M. Leung, Photonic band structure: the face-centered-cubic case employing nonspherical atoms, Phys. Rev. Lett. **67**(17), 2295–2298 (1991).

[42] P. Schmuki, L.E. Erickson, D.J. Lockwood, J.W. Fraser, G. Champion and H.J. Labbe, Formation of visible light emitting porous GaAs micropatterns, Appl. Phys. Lett. **72**, 1039–1041 (1998).

[43] M.N. Ruberto, X. Zhang, R. Scarmozzino, A.E. Willner, D.V. Podlesnik and R.M. Osgood, The laser-controlled micrometer-scale photoelectrochemical etching of III–V semiconductors, J. Electrochem. Soc. **138**, 1174–1187 (1991).

[44] X.G. Zhang, Mechanism of pore formation on n silicon, J. Electrochem. Soc. **138**, 3750–3756 (1991).

[45] H. Gerischer, P. Allongue and V. Costa Kieling, The mechanism of the anodic oxidation of silicon in acidic fluoride solutions revisited, Ber. Bunsenges. Phys. Chem. **97**, 753–757 (1993).

[46] J. Carstensen, M. Christophersen and H. Föll, Pore formation mechanisms for the Si–HF system, Mater. Sci. Eng. B **69–70**, 23–28 (2000).

[47] G.S. Popkirov and R.N. Schindler, New approach to the problem of "good" and "bad" impedance data in electrochemical impedance spectroscopy, Electrochim. Acta **39**, 2025–2030 (1994).

[48] C. Jäger, B. Finkenberger, W. Jäger, M. Christophersen, J. Carstensen and H. Föll, Transmission electron microscopy investigation of the formation of macropores in n- and p-Si(001)/(111), Mater. Sci. Eng. B **69–70**, 199–204 (2000)

[49] Z.H. Lu, F. Chatenoud, M.M. Dion, M.J. Graham, H.E. Ruda, I. Koutzarov, Q. Liu, C.E.J. Mitchell, I.G. Hil and A.B. McLean, Passivation of GaAs(111)A surface by Cl termination, Appl. Phys. Lett. **67**, 670–672 (1995).

[50] F. Müller, A. Birner, J. Schilling, U. Gösele, Ch. Kettner and P. Hänggi, Membranes for micropumps from macroporous silicon, Phys. Status Solidi a **182**, 585–590 (2000).

[51] J. Carstensen, R. Prange, G.S. Popkirov and H. Föll, A model of current oscillations at the Si-HF-system based on a quantitative analysis of current transients, Appl. Phys. A **67–4**, 459–467 (1998).

[52] J. Carstensen, R. Prange and H. Föll, Percolation model for the current oscillation in the Si–HF system, *Proc. ECS' 193rd Meeting*, San Diego, Vol. 98-10, 1998, pp. 148–157.

[53] M. Christophersen, J. Carstensen, A. Feuerhake and H. Föll, Crystal orientation and electrolyte dependence for macropore nucleation and stable growth on p-type Si, Mater. Sci. Eng. B **69–70**, 194–198 (2000).

5

Microporous Honeycomb-Structured Polymer Films

L.V. Govor

Institute of Physics, University of Oldenburg, D-26111 Oldenburg, Germany
leonid.govor@uni-oldenburg.de

5.1. INTRODUCTION

Micrometre-sized porous membranes with highly ordered honeycomb structures are technologically important for a variety of applications. They are of interest for use in chemistry and life science [1]. As a consequence of their strong periodicity, these membranes made from diverse compounds possess the characteristic properties of polymer photonic crystals [2–4]. For the manufacturing of such membranes, a large variety of the methods exists [5]. Self-organization may offer some advantages in this field: the patterning process can function in different media, and cost-intensive large-scale technology is not necessary. By self-organized processes, one is able to develop distinct structures with a regular geometrical configuration (e.g., networks with hexagonal cells). Nowadays, the ordered mesoporous solids with nanoscale pore sizes are fabricated by self-organization of spherical micelles from a diblock copolymer system in a selective solvent [6]. The mesoporous materials can be formed by colloidal templating [7,8]; colloidal crystals of polystyrene or silica spheres are composed of fluid that fills the space between the spheres. Thereafter, the templating spheres are removed to result in the creation of a porous solid where the dimension of the pores matches those of the templating spheres.

Water-assisted formation of ordered mesoporous membranes was described in the numerous articles [9–14]. In this case, the ordered membranes are formed by the condensation of water vapour on the fluid polymer solution film and by a subsequent evaporation of a solvent from the polymer solution. Recently, we have developed a preparation method that allows the membrane formation with evenly shaped hexagonal cells with a diameter of about 1–2 μm [15–17]. In the above works, we have described in detail the technology how to get mesoporous membranes from different polymers. Simultaneously, some effort has been undertaken to interpret the self-organizing mechanism of patterning by thermo-dynamic processes that take place between water droplets on the fluid polymer solution surface. We have suggested that the stabilization of water droplets on a fluid surface is indispensable for ordered structure formation. In the following, the most important stabilization parameters are discussed that can influence the growth of condensing water droplets on the fluid polymer solution layer and their interaction between each other.

5.2. EXPERIMENTAL FORMATION OF POLYMER HONEYCOMB STRUCTURES

For the formation of a self-organized honeycomb polymer network, we have de-veloped a four-step method. Figure 5.1 shows these steps: (a) deposition of one drop of polymer solution (liquid F1) on the cooled water surface (liquid F2, 3–5 °C); (b) spread-ing of one drop of polymer solution to an extremely thin layer; (c) interaction of water vapour (air with 100%, 75%, 32% or 19% relative humidity at 20 °C) with the polymer thin film surface for structuring/self-organization of the lateral water droplets distribution and polymer network formation and (d) mechanical removal of the structured network from the liquid and potential fixation by annealing.

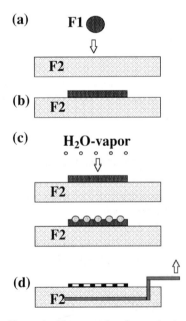

FIGURE 5.1. Formation of a self-organized honeycomb polymer structure: (a) deposition of one drop of polymer solution F1 on the cooled water surface F2; (b) spreading of one drop of polymer solution to a thin layer; (c) water vapour condensing on the polymer film surface; growing of water droplets and building of the compact hexagonal structure, i.e., polymer network and (d) drying of the polymer network and transfer from the water surface to a fixed substrate.

The condition for the spreading of a drop of liquid F1 on the surface of liquid F2, is given by thermodynamic arguments [18]

$$\sigma_{F2/G} > \sigma_{F1/G} + \sigma_{F1/F2} \qquad (5.1)$$

where $\sigma_{F1/F2}$, $\sigma_{F1/G}$ and $\sigma_{F2/G}$ denote the surface tension between the liquids F1 and F2, between liquid F1 and gas G (G is air) and between liquid F2 and gas G, respectively. The degree of spreading of a film of liquid F1 over liquid F2 is characterized by the spreading coefficient which is determined as [18]

$$S_{F1/F2} = \sigma_{F2/G} - \sigma_{F1/G} - \sigma_{F1/F2}. \qquad (5.2)$$

For $S_{F1/F2} > 0$, the total spreading is achieved, i.e., the liquid drop F1 will cover the whole surface of the liquid F2 and, thereby, forms a monomolecular layer at the edge. For $S_{F1/F2} < 0$, there is no spreading.

5.2.1. Formation of Nitrocellulose Networks

The coating of one drop of the 1% nitrocellulose solution in amyl acetate on the cooled water surface leads to a complete spreading along that surface [16]. The surface tension coefficient of the 1% nitrocellulose solution in amyl acetate has a value $\sigma_{F1/G}$ (20 °C) = 24.6 mN/m. Distilled water which was cooled down to 3–5 °C possesses a surface tension coefficient of $\sigma_{F2/G}$ = 74.9 mN/m. For the surface tension of the 1% nitrocellulose solution in amyl acetate on water, we take the value $\sigma_{F1/F2}$ = 12 mN/m. The demand for the total spreading of one drop of the 1% nitrocellulose solution in amyl acetate on the cooled water surface is fulfilled because we have $S_{F1/F2}$ = 38.3 mN/m.

The size of the spread thin polymer layer in the vessel (with a diameter of 93 mm) was 70 mm. Since the volume of the spread drop was 0.015 cm³, the thickness of the resulting spread liquid polymer layer can be estimated to 3.9 μm. Our thin film was subject to the influence of water vapour which induces the self-organized formation of a honeycomb network structure. There was no control of the size of the water droplets, but there was control of the relative humidity of air which was about 75% at a temperature of 20 °C. Depending on the time elapsed after the water vapour has begun to affect the polymer film, one obtains a variety of network distinguished both in form and size. During our experiments, the above time span changed between 1 second and 60 seconds. In the final step, after having dried the network, it was transferred from the surface of water to a sapphire substrate.

5.2.2. Formation of Poly(p-Phenylenevinylene) and of Poly(3-Octylthiophene) Networks

For the formation of a self-organized honeycomb poly(p-phenylenevinylene) (PPV) and poly(3-octylthiophene) (P3OT) networks, we have used (a) a 2% xylene solution of the PPV precursor with $\sigma_{F1/G}$(20 °C) = 30 mN/m and (b) a 2% xylene solution of polythiophene with a nearly equivalent value of the surface tension. The surface tension between both 2% polymers solution in xylene and water (liquid F2) amounts to $\sigma_{F1/F2}$(20 °C) = 36.1 mN/m. From Equation (5.2), we can calculate the spreading coefficient that is about $S_{F1/F2}$ = 8.8 mN/m, i.e., the condition (1) is fulfilled.

The thin polymer solution film formed on the surface of liquid water was then (immediately after spreading) subjected to water vapour (air with 19% relative humidity at a temperature of 20 °C). After the self-organization process of the water droplets on the polymer film has taken place, the polymer film builds up a honeycomb structure. It should be noted that the diameter of the initial spread liquid polymer layer in the vessel (with a diameter of 93 mm) at the beginning was about 50 mm. Then the size of liquid polymer layer decreases to about 20–30 mm. The thickness of the resulting dry polymer layer was not homogeneous. The edge of the polymer layer was thinner and in this area the formation of hexagonal network occurs. The middle area of the layer was thicker and not structured. After having dried the network, it was transferred from the surface of water to a substrate (quartz, glass or indium–tin–oxide-plated glass).

5.3. SELF-ASSEMBLED NETWORKS OF POLYMERS

5.3.1. Nitrocellulose Networks

In the following, we will consider the network structures, obtained by the procedure described in Section 5.2. Having brought the water vapour on the polymer layer and having dried the polymer layer with a diameter of 70 mm, a fractal-like geometry occurred, i.e., areas (stripes) with network structures and areas without them are found. The majority of the structured stripes was distributed at the edge of the polymer layer and had normally the width of 0.5 mm and the length of 20 mm. The stripes can be connected with or separated from each other. Figure 5.2a shows a scanning electron microscope (SEM) picture of hexagonal nitrocellulose cells as a fragment. In this case, the water vapour was coated 10 seconds after spreading the polymer layer onto the cooled water surface. Note the strong homogeneity and reproducibility of the individual cells inside the network structure. For a more detailed look at the geometry, the picture of only a few cells is displayed in Figure 5.2b. The interpore distance amounts to 2.6 μm and the width of the walls to about 0.4 μm. The cross section of the pore wall perpendicular to the photograph plane offers the shape of a T-section. The latter means that each hexagonal cell lies on a hexagonal base also having extremely thin side walls. The height of the base is 0.5 μm. Each one of these six side walls represents some kind of a frame with various thicknesses along the circuit on which a thin polymer film is stretched. In most cases, the polymer film unveils an oval aperture in the middle of the frame.

Few structured big stripes have a different interpore distance on the edge (for example, 2 μm) and at the centre (for example, 6 μm) of the stripe. Figure 5.3 shows a SEM image of different fragments of one structured stripe after the annealing for 1 hour at a temperature of 950 °C under vacuum conditions. Figure 5.3a shows the network fragment between the edge and the centre of the stripe. It can be seen that the cross section of the pore wall perpendicular to the photograph plane for a cell with the diameter about 2–4 μm does not represent the shape of a T-section, i.e., the pore wall merges together after annealing. As a result, we observe two-dimensional cells. Figure 5.3b shows the network fragment at the centre of stripe after annealing. A three-dimensional cell from a carbon network pattern obtained from a corresponding nitrocellulose structure after

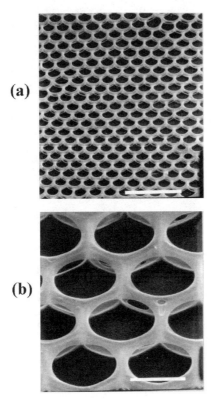

FIGURE 5.2. SEM images of a nitrocellulose network which were prepared via coating the water vapour 10 seconds after having spread the polymer layer onto the cooled water surface. Scale bars: (a) 10 μm and (b) 2 μm.

annealing is shown in Figure 5.3b. Regions of such structural form extend over an area of about (30 × 30) μm². Note that such kind of network patterning was only observed in the case where the cells have a relatively large diameter. It is obvious that the upper and lower hexagonal cells (one placed on the top of the other)—both with a diameter of 6 μm—are practically equal. The two cells are connected only at the corners. The height of the connection is approximately 1.5 μm. The diameter of the pore wall amounts to approximately 0.25 μm.

Figure 5.4a gives another SEM images of a nitrocellulose network with a different kind of structure (in contrast to Figure 5.3) that we have obtained when coating the water vapour 60 seconds after spreading the polymer layer onto the cooled water surface. Take note of the parallel orientation (in three directions) of the pores. For a better determination of the geometrical parameters, a few cells are shown in Figure 5.4b. It turns out that the cross section of the pore walls is plate shaped. The width of the pore walls is approximately 1.5 μm and the height of the "plate walls" about 0.25 μm. Inside the cell, one can perceive the breakthrough of the thin polymer film. Take note of the fact that the cross section of the pore walls of the latter structure does not have the shape of a T-section (in contrast to Figure 5.2).

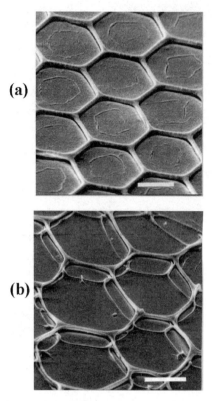

(a)

(b)

FIGURE 5.3. SEM images of different areas of one network with (a) a two-dimensional single cell and (b) a three-dimensional cell which was heated for 1 hour at a temperature of 950 °C under vacuum conditions. The network was prepared via coating the water vapour 10 seconds after having spread the polymer layer onto the cooled water surface. Scale bars: (a) 2 μm and (b) 4 μm.

5.3.2. Poly(p-Phenylenevinylene) and Poly(3-Octylthiophene) Networks

Figures 5.5 and 5.6 show SEM images of a fragment of the PPV precursor and polythiophene networks taken at different magnifications. The strong periodicity and homogeneity of the individual cells can be clearly recognized. For a closer look at the profile of the networks, atomic force microscopy (AFM) studies have been performed. The contact-mode topographic profiles of the PPV precursor network are displayed in Figure 5.7. The depth of the pores is 0.6 μm, the diameter about 1 μm and the width of the pore walls 0.7 μm. Corresponding values for conjugated polythiophene are 0.5 μm, 1.0 μm and 0.8 μm, respectively. Note that the pore wall of the PPV network is substantial in contrast to the polythiophene wall which is similar as the cellulose wall in Figure 5.4b. To end up with a conjugated PPV network, the sample of the PPV precursor network was annealed for 2 hours at a temperature of 160 °C under nitrogen flow atmosphere. The main finding is the following: the PPV precursor network converts to a conjugated PPV film that consists of periodic hill-like structures with a translational periodicity of the precursor network, i.e., the network melts to the film.

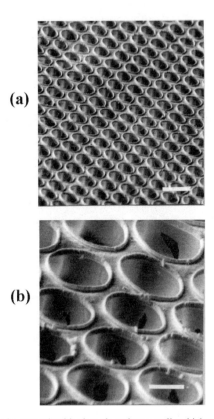

(a)

(b)

FIGURE 5.4. SEM images of a network with plate-shaped pore walls which were prepared via coating the water vapour 60 seconds after having spread the polymer layer onto the cooled water surface. Scale bars: (a) 5 μm and (b) 2 μm.

5.4. MODEL FOR THE FORMATION OF THE HONEYCOMB STRUCTURES IN POLYMER FILMS

When the fluid substance is placed at a liquid–air interface, it may spread out to thin film. It happens when a liquid of low surface tension is placed on one of a high surface tension. We have a positive spreading coefficient for all our polymer solutions in amyl acetate and in xylene. We have suggested that the size of thin xylene polymer solution layer on the water surface decreases with time. The latter can be explained as follows. In the experimental part, for the determination of the spreading coefficient $S_{F1/F2}$ (Equation (5.2)) we have used the surface tension values of the xylene solution, amyl acetate solution and water as for the pure liquids. However, when two liquids are in contact, they will become mutually saturated, so that $\sigma_{F2/G}$ will change to $\sigma_{F2(F1)/G}$, and $\sigma_{F1/G}$ to $\sigma_{F1(F2)/G}$. The symbol F1(F2) means that the liquid F1 is saturated with the liquid F2. The corresponding spreading coefficient has the symbol $S_{F1(F2)/F2(F1)}$ and can be determine as [18]

$$S_{F1(F2)/F2(F1)} = \sigma_{F2(F1)/G} - \sigma_{F1(F2)/G} - \sigma_{F1/F2}. \tag{5.3}$$

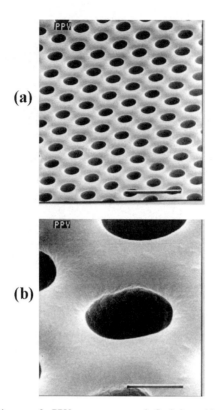

FIGURE 5.5. SEM images of a PPV precursor network. Scale bars: (a) 6 μm and (b) 1 μm.

In this case, for the solution of nitrocellulose in amyl acetate on the water surface (at 20 °C), we have $S_{F1(F2)/F2(F1)} = 39.5\,\text{mN/m} - 26.9\,\text{mN/m} - 12\,\text{mN/m} = 0.6\,\text{mN/m}$. The corresponding spreading coefficient for solutions of PPV (or P3OT) in xylene on the water surface is $S_{F1(F2)/F2(F1)} = 55.7\,\text{mN/m} - 28.8\,\text{mN/m} - 36.1\,\text{mN/m} = -9.2\,\text{mN/m}$, i.e., the final spreading coefficient is negative and the xylene polymer solution film retracts to a lens. This retraction is substantially smaller for the solution of nitrocellulose in amyl acetate. The values of the surface tension were determined using a stalagmometer. This instrument consists of a capillary tube through which the polymer solution flows, which enables the counting of the number of droplets and, therefore, the derivation of the surface tension.

Next, we look in a more detailed way at the lay on of the steam, because it represents one crucial factor for the patterning process. Since the cell diameter of our networks is about 2 μm, we investigate the shape of the water droplet with a diameter of 2 μm, before it dissolves on the surface of the polymer thin film. We have to compare two pressure quantities: the first one is the capillary pressure P_L which gives rise to the spherical shape of the droplet, the second one, the so-called gravitation pressure P_G, causes the flattening of the droplet. The capillary pressure can be calculated by the well-known formula $P_L = 2\sigma_{F2/G}/R$, with R giving the droplet radius. In our case, we obtain $P_L = 1.4 \times 10^9$ Pa. If the contact plane between the droplet and the polymer thin film

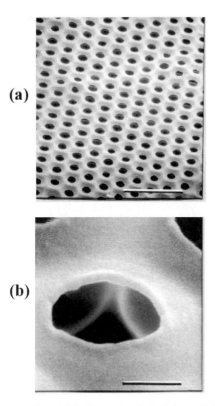

FIGURE 5.6. SEM images of a polythiophene network. Scale bars: (a) 6 μm and (b) 0.5 μm.

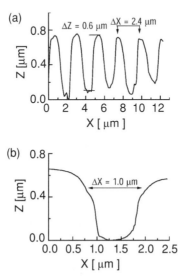

FIGURE 5.7. Contact-mode AFM images of the PPV precursor: (a) the profile of several pores to demonstrate the height distribution and (b) the profile of a single pore. Z: height; X: lateral coordinate.

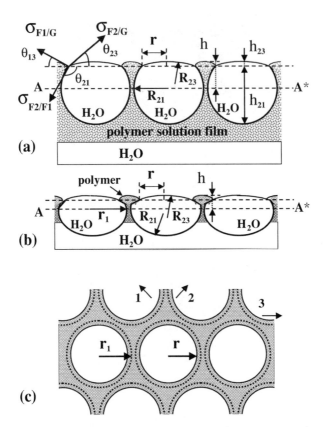

FIGURE 5.8. Model for the lay on of the water steam on the polymer layer: (a) process of envelopment of the water droplet by the polymer layer; the thin film on the surface of the water drop indicates a monomolecular polymer layer; (b) moment of the first contact of the water drop with the cooled water surface and (c) model structure of the network according to (b) in the plane of the polymer layer.

is approximately πR^2, we can determine $P_G = mg/\pi R^2$ with the water droplet mass $m = 4\pi R^3 \rho/3$ (ρ is the water density). Finally, one ends up with $P_G = 1.3 \times 10^{-2}$ Pa. If we compare both types of pressure, it is obviously $P_L \gg P_G$. This means that the water droplet has a spherical shape before laying on the surface of polymer solution. The first moment the water droplet contacts the polymer layer is sketched in Figure 5.8a: As the droplet touches the surface, its destruction will start. The angles θ_{21} and θ_{23}, in this case, are nonequalized boundary angles, because they experience variations in the development of further physical and chemical processes that take place at the phase boundary.

Water is a denser medium than amyl acetate (0.87 g/cm³) and xylene (0.87 g/cm³), but the water droplets do not sink because of a subtle balance between buoyancy, droplet weight and capillary forces [19]. The droplets are situated at the interface between the fluid polymer layer and air. Only a small part of the droplet is situated above the surface of the fluid polymer layer. The water droplets on the fluid polymer layer do not coalesce immediately when they touch because they are separated by a thin film of the polymer solution. It is assumed that a thin polymer film (only a few monomolecular layers thick) is formed on the top of the droplets similar to the spreading of the polymer solution drop on the water surface. This thin film on the top of the water droplet might be the cause for stopping the growing of the water droplets and therefore a key parameter for the droplets size regulation. Additionally, we have suggested that water droplets in the polymer solution are covered with a solid polymer layer at the interface between the two liquids. Experimentally, this layer can be seen in Figures 5.4 and 5.6, which are characterized by the bursting holes (black); the water flows out of them after the solvent has evaporated. This layer prevented coalescence of water droplets, i.e., the precipitation of nitrocellulose, PPV and P3OT at the interface between the polymer solution and the water droplets is the basis for the formation of a compact hexagonal structure of water droplets. In our modelled Figure 5.8a, the precipitation layer is illustrated as a fat line between the water droplets and the polymer solution. Similar results were determined for the system of water droplets on the surface of solution of poly(p-phenylene)-block-polystyrene in carbon disulfide [12]. In Figure 5.8a, the corresponding forces which come into play at the above phase boundary are indicated. These forces converted to the length unit of the wetting line are equivalent to the corresponding quantities of the surface tension. From the condition that gives the balance of the surface tensions, the equilibrium state at the edge circumference of the contact between the water droplet and the polymer layer can be described by the following two equations

$$\sigma_{F1/G} \cos \theta_{13} = \sigma_{F2/G} \cos \theta_{23} + \sigma_{F2/F1} \cos \theta_{21}, \tag{5.4}$$

$$\sigma_{F2/G} \sin \theta_{23} + \sigma_{F1/G} \sin \theta_{13} = \sigma_{F2/F1} \sin \theta_{21}, \tag{5.5}$$

with the angles θ_{ij} as defined in Figure 5.8a. Equation (5.4) treats the balance of forces in the plane of the polymer layer and Equation (5.5) the balance of forces directed perpendicular to it.

For further qualitative discussion of the water droplet shape (the angles θ_{ij} in Figure 5.8a) only the initial values of surface tension σ had been used. The water droplet shape on the surface of paraffin oil had been studied by Knobler and Beysens [20] and on the surface of carbon disulfide by Pitois and Francois [12]. They have suggested that the water droplet has the form of a very asymmetric lens, the major part of which is nearly a complete sphere suspended from the surface. Based on these studies and from our experimental results in Figures 5.4–5.6, the shape of the water droplet on the surface of the polymer solution in xylene (or amyl acetate) has been sketched in Figure 5.8a. In our case, the capillary pressure in water droplet is drastically higher as the gravitation pressure ($P_L \gg P_G$), i.e., the influence of the gravity forces on the droplet shape can be neglected. This means that the small water droplet on the fluid polymer layer is composed of two spherical segments, whose angles θ_{ij} and radii R_{ij} can be determined by the surface tension forces [21]:

$$\cos \theta_{ij} = [1 + (\sigma_{ij}/\sigma_{jk})^2 - (\sigma_{ik}/\sigma_{jk})^2]/2(\sigma_{ij}/\sigma_{jk}), \tag{5.6}$$

$$R_{ij}/r = 1/\sin \theta_{ij}, \tag{5.7}$$

$$h_{ij}/r = (1 - \cos \theta_{ij})/\sin \theta_{ij}, \tag{5.8}$$

where θ_{ij}, R_{ij}, h_{ij} and r are defined as illustrated in Figure 5.8a. It should be noted that in this case, for an extremely small lens, Princen [21] has assumed that angle

θ_{13} in Figure 5.8a is zero. From Equation (5.6), for the water droplet on the surface of the polymer solution in xylene with $\sigma_{F1/G} = 30.0$ mN/m, $\sigma_{F2/F1} = 55.7$ mN/m and $\sigma_{F2/G} = 72.9$ mN/m, we have determined the angles which are $\theta_{21} = 113°$ and $\theta_{23} = 45°$. On the other hand, for the water droplet on the surface of paraffin oil, the angles θ_{ij} were measured by Knobler and Beysens [20]. From this study it follows that the values of the angles are $\theta_{13} = 135°-140°$, $\theta_{23} = 20°-25°$. The fluid properties (surface tension, density) of xylene and of paraffin oil are comparable. The force $\sigma_{F1/G} \sin \theta_{13} \times 2 \pi R$ in Equation (5.5) does not permit that the water droplet sinks in the polymer solution layer, i.e., the angle θ_{13} cannot be 0. For the water droplet with a diameter of 2 μm in our case, the deflection of the force $\sigma_{F1/G}$ from a horizontal line is not very large and a experimental determination of the angle θ_{13} is not easy.

At the moment of the first contact between the water droplet and the fluid polymer layer the force $\sigma_{F2/G} \sin \theta_{23}$ for achieving the equilibrium state (described by Equation (5.5)) tends to pull the polymer layer onto the water droplet. The water droplet more and more penetrates the polymer layer. The envelopment of the water droplet by the polymer layer will take place as long as the angles θ_{21}, θ_{23} and θ_{13} (or the diameter of the surface curvatures, R_{21} and R_{23}, see Figure 5.8a) experience variations such that the pressure inside the water droplet caused by the surface curvature of the boundary water—air is equal to the pressure which results from the surface curvature at the boundary water—polymer layer. Their balance can be expressed mathematically by

$$\sigma_{F2/G}/R_{23} = \sigma_{F2/F1}/R_{21}, \tag{5.9}$$

where R_{23} and R_{21} are the radii of the water droplet at the boundary water—air and water—polymer layer, respectively. If we substitute the above values in Equation (5.9), we obtain the ratio $R_{23}/R_{21} \approx 1.3$ for the polymer solutions in xylene. This means that pressure equality in the lower and upper part of the water droplet at thermodynamical equilibrium can only be granted if R_{23} exceeds R_{21} by a factor 1.3 (see Figure 5.8a) for PPV (or P3OT) in xylene. On the other hand, a same value of the relation R_{23}/R_{21} can be determined from Equation (5.7), i.e., $R_{23}/R_{21} = \sin \theta_{21}/\sin \theta_{23} \approx 1.3$ with $\theta_{21} = 113°$ and $\theta_{23} = 45°$.

The formation of the hexagonally arranged layer of the water droplets on the liquid polymer layer can be explained as follows. With the water vapour condensation on the cold surface of the liquid, the water droplets form to a pattern of breath figures [22–25], for which the geometry can be very different. The physical basis of the breath figure formation on the fluid surfaces is discussed by Knobler and Beysens [20] and by Steyer et al. [26,9]. It is suggested that the formation of breath figures on fluid surfaces evolves through three stages: (a) initial stage, when droplets are isolated and do not interact strongly, and the average droplet radius $\langle R \rangle$ increases with time as $\langle R \rangle \propto t^{1/3}$; (b) crossover stage, when the surface coverage is high, and the rate of the droplet growth increases and (c) coalescence-dominated stage, when the surface coverage is high and constant, and the droplet radius increases as $\langle R \rangle \propto t$.

The attractive force F between two water droplets on the surface of the polymer solution separated by a distance l is determined by Chan et al. [19] and by Steyer et al. [9] as

$$F = (4\pi R^6 \rho^2 g^2/3l\sigma_{F1/G})[1/\rho_s + 0.25(1 - p^2)^{1.5} - 0.75(1 - p^2)^{0.5}]^2 \tag{5.10}$$

where R is the radius of the droplets, ρ is the absolute density of the polymer solution, g is the earth's gravitational acceleration, $\sigma_{F1/G}$ is the surface tension between the polymer solution and air, ρ_s is the relative density of the polymer solution to air and $p = r/R_{21}$, where r and R_{21} are radii illustrated in Figure 5.8a. This equation is valid when the bond number $B_0 = R^2 \rho g / \sigma_{F1/G}$ is small enough (<0.1, see [19]). In our case, B_0 is of the order of 10^{-6} for all the polymer solutions and for the water droplets with a diameter of 2 μm. The parameter p characterizes the contact angles of the water droplet with the surface of the polymer solution and is very sensitive to the wetting properties between the water and the polymer solution. Steyer et al. [9] have suggested that the formation of the hexagonal structure of water droplets occurs when the parameter p has a value in the range from 0.4 to 0.8. The hexagonal structure disappears when p has a value in the range from 0.2 to 0.4.

In our case, the exact determination of parameter p and the corresponding force F is not easy, because initial values of the parameters $\sigma_{F2/G}$, $\sigma_{F1/G}$, $\sigma_{F2/F1}$ and ρ alter with time. In this case, it is of interest to consider only some tendency according to Equation (5.10) with initial values of the surface tension σ. From Equation (5.10) follows that by altering p the force can increase considerably, i.e., the value of p can be a key parameter for the formation of a hexagonal structure of water droplets. In our case, the value of the parameter p can be estimated from Equation (5.7). For the value of the angle $\theta_{21} = 113°$, the calculated value is $p \approx 0.9$. The corresponding experimental value of p which follows from geometrical sizes of a single cell in Figures 5.2 and 5.5 is equal to 0.8 for a nitrocellulose network and 0.6 for a P3OT network. From Equation (5.8) follows that for $\theta_{23} = 45°$ and $r \approx 0.9$ μm (for $R_{21} = 1$ μm) the value of the parameter $h_{23} \approx 0.4$ μm and for $\theta_{21} = 113°$ the value of the parameter $h_{21} \approx 1.4$ μm.

On increasing $\sigma_{F1/G}$, the attractive force F between the two water droplets decreases and the corresponding distance l between droplets increases. We can see this fact in Figures 5.2 and 5.5. The width of the pore walls for the solutions of PPV in xylene (with $\sigma_{F1/G}$ (20 °C) = 30 mN/m) is 0.7 μm, and for a nitrocellulose solution in amyl acetate (with $\sigma_{F1/G}$ (20 °C) = 24.6 mN/m) the width is 0.25 μm. Turning to Figure 5.4, the corresponding nitrocellulose network which is obtained when coating the water vapour 60 seconds after spreading the polymer layer onto the cooled water surface has the width of the pore wall about 1.0 μm. This means that the surface tension of nitrocellulose solution increased (with solvent evaporation).

During the drying process of the polymer film, the water droplets approach the cooled water surface. In Figure 5.8b, the contact between the water droplet and the cooled water surface together with the corresponding variation in the radii R_{23} and R_{21} are displayed. Between the water droplet and the cooled water surface, we find a thin precipitation polymer film. The radius R_{21} will be enlarged, since the lower part of the water droplet which is in contact with the cooled water surface (more precisely, with the thin polymer film) will have a flat boundary plane. In Figure 5.8b, the cross section of the pore wall of one cell of the polymer network is shown which develops between two neighbouring water droplets. If we compare Figure 5.8b and Figure 5.2, we can recognize a pronounced similarity between the cross section of the experimentally prepared single cell of the polymer network and the one of the above modelled cell. From the figures, we conclude that the minimum thickness of the polymer layer in the middle of the frame (i.e., the base plate) corresponds to the minimum distance between the water droplets near the AA* line. For nitrocellulose in amyl acetate, this distance is probably a few monomolecular layers thick, but for PPV in xylene this distance is about 0.7 μm. It should be noted that this distance difference between nitrocellulose and PPV can be explained first of all with the larger distance between condensed water droplets on the surface of the solution PPV in xylene. It follows from Equation (5.10) that increasing $\sigma_{Fl/G}$ for PPV in xylene in comparison to nitrocellulose in amyl acetate leads to a smaller attractive force F between two water droplets and, accordingly, to a larger distance l between droplets. Second, it is possible that the precipitation of PPV at the solution–water interface is larger than that of nitrocellulose.

The appearance of small areas ((30×30) μm^2) at the centre of the nitrocellulose network stripes (Figure 5.2, the single cell on the edge of stripe is of 2 μm), with basic cells as displayed in Figure 5.3b, results from some size gradient of the water vapour droplets. The formation of hexagonally arranged layer of water droplets on the liquid polymer layer with a size gradient of water droplets can be explained as follows. With the water vapour condensation on the cold surface of the liquid in the first moment, the water droplets conform the islands which build later the breath figures [22–25]. The islands attract each other in a much stronger way than the single droplets. This is because the attractive force F between the islands is proportional to the sixth power of their radii as in Equation (5.10). The intrinsic droplets in the stripe have the maximal shrinkage, and they can coalesce when a polymer film between the water droplets is not large enough. In the case of the water vapour with an extension comparable to the thickness of the polymer layer (4 μm), the total polymer solution will even move to the space in between the water droplets. As a consequence, there only remains a quite thin polymer film underneath the water droplets such that the polymer solution will be uniformly distributed with respect to the AA* line. The subsequent evaporation of amyl acetate out of the solution does not give rise to an essential redistribution of the polymer relating to the AA* line. We end up with the result that the cells shown in Figure 5.3b are, with respect to the third dimension, deeper (1.5 μm) than those shown in Figure 5.2 (0.5 μm). Moreover, they undergo a stronger symmetry with respect to the AA* line.

Figures 5.2, 5.4–5.6 and 5.8 give rise to the assumption that, first, the solidification of the polymer network (i.e., the evaporation of amyl acetate or xylene) takes place, and afterwards the water droplet will break through the thin polymer film and flow into the cooled water (compare the bursting holes (black) within the cells in Figures 5.4 and 5.6). As already noted above, the thickness of the liquid polymer layer is about 4 μm after dissolving on the water surface. During the drying process, the thickness of the polymer layer amounts to about 0.5 μm. As you can see in Figures 5.2 and 5.8, the whole polymer dissolved in amyl acetate (or xylene) is solidified in the AA* region, more precisely, above that line.

If the height h (Figure 5.8b) of the enveloping polymer layer is not too large, a hexagonal patterning can be achieved. The experimental realization is illustrated in Figures 5.2, 5.5 and 5.6. The underlying basic structure is outlined in Figure 5.8c. This means that, at a low height, the width of the pore wall will also be small. The junctions cross in the knots of the hexagonal network at the angle of 120° and usually form a continuous network. In Figure 5.8c, you can see that the hexagonal network developed at small values of h results in three types of zig-zag lines which proceed to the directions

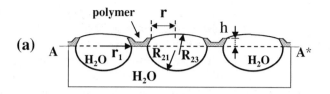

(a)

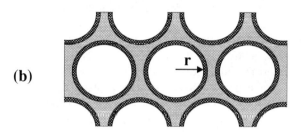

(b)

FIGURE 5.9. Model structure of the network according to the experimental situation in Figure 5.4. (a) Shape of the cross section of the pore wall and (b) form of the network. Parameter r_1 is the radius of the deformed water droplets.

1, 2, 3. It must be emphasized that the hexagonal shape of the single cells of the polymer network becomes clearer during evaporation.

In the case where the water vapour begins to affect the nitrocellulose layer not until 60 seconds after spreading it onto the cooled water surface, the thin polymer film gradually becomes dry in the meantime. In other words, the evaporation of amyl acetate out of the polymer layer during the elapsed time of 60 seconds gives rise to an increasing concentration of the polymer inside the polymer film. As a consequence, the distance between the neighbouring droplets of water vapour that precipitate onto the liquid surface of the polymer layer increases ($\sigma_{F1/G}$ is larger, Equation (5.10)). Accordingly, the same holds for the interspace between the neighbouring droplets of the water vapour where the polymer material accumulates after the polymer thin film has completely become dry (in the vicinity of the AA* line). In this case, the cross section of the pore wall will have the shape of a plate (Figure 5.9a), and the whole network resembles that one shown in Figure 5.9b. The lifted edges in Figure 5.9a are displayed together with the different pattern in Figure 5.9b. Upon comparing the experimental results in Figure 5.4 with the modelled networks in Figure 5.9, one recognizes a good correspondence.

5.5. APPLICATION OF POLYMER NETWORKS

5.5.1. Nitrocellulose Networks as Precursor for Carbon Networks

Carbon networks fabricated by the above-described method of a self-organized process represent porous disordered systems for electrical conductivity. The degree of disorder and, accordingly, the values of their electrical conductivity extending from insulator to metal behaviour change via heat treatment under vacuum conditions at process temperatures in the range from 600 °C to 1000 °C. Upon varying the ambient temperature from 4.2 K to 295 K, four transport mechanisms can be observed [27,28]. For carbon networks (Figure 5.3a) whose conductivity is far beyond the metal–insulator transition (MIT), the specific resistivity ρ_c depends on the temperature T as $\rho_c(T) \propto T^{-b}$ exp $([T_0/T])^{1/m}$, where T_0 is a constant. In the low-temperature range, a Coulomb gap in the density of states located near the Fermi energy level occurs, that means, the characteristic value of the exponent is $m = 2$. At high temperatures, the pre-exponential part $\rho_c(T) \propto T^{-b}$ dominates. In the intermediate temperature range, Mott's hopping law with $m = 3$ follows. However, the specific resistivity of the carbon networks subject close to the MIT follows the power law $\rho_c(T) \propto T^{-b}$ with $0 \leq b \leq 3$ at low temperatures. In the high-temperature range, the specific resistivity is characterized by $\rho_c(T) \propto$ $\exp(-[T/T_1]^{c-1})$, where the values for c are varying from 1.3 to 1.5 and T_1 is a constant. The above four charge transport mechanisms can be explained by the tails in the density of localized states pulled out of the conduction and valence band, as a consequence of disorder and in particular by some overlap between these tails.

The influence of the electrical field on the variable range hopping process of porous carbon networks is examined in the range of the validity of the law $\ln \sigma_c(T) \propto -(T_0/T)^{1/2}$ (Coulomb gap), where σ_c means electrical conductivity. It is shown that the field dependence of the samples investigated in the vicinity of the metal–insulator transition clearly distinguishes four characteristic regions [29]. At low values of the electrical field applied, we have ohmic conductivity. Upon increasing the electrical field E, the electrical conductivity σ_c rises, first following the law $\ln \sigma_c(E) \propto E^n$, where n changes from 1.4 to 2.6 with increasing distance from the metal–insulator transition on the insulating side. Then, at higher electrical field, the conductivity turns to the relation $\ln \sigma_c(E) \propto E^{1.0}$. The temperature dependence of the hopping length l_h of the charge carriers, determined within the above field regime, develops as $l_h(T) \propto T^{-0.9}$. At temperatures where the ohmic behaviour in the Coulomb gap occurs and obeys the law $\ln \sigma_c(T) \propto -(T_0/T)^{1/2}$, the electrical conductivity caused by thermally nonactivated charge carriers at high fields complies with $\ln \sigma_c(E) \propto E^{-1/3}$. The current density j changes as $\ln j(E) \propto E^{-1/6}$. The temperature dependence of the threshold electrical field E_{th}, which characterizes the transition from the low-field to the high-field range, follows $E_{th} \propto T^{1.5}$.

5.5.2. Nitrocellulose Network as Mask for Ion-Etching Process

The nitrocellulose network is placed on a GaAs epitaxial layer with an area of (240 × 330) μm^2, a thickness of 0.25 μm and an electron concentration of $1.3 \times 10^{17} cm^{-3}$ (Figure 5.10a). On this layer, GeAu contacts ensure ohmic contacts. The network represents a mask for the following Ar-ion-etching process with an energy of 3.5 keV. Due the difference in the etching velocity between cellulose and GaAs, the cellulose mask was totally removed in a time span of 90 seconds. Also, the GaAs layer is removed in the meshes of the cellulose network, and the network structure is transferred to the GaAs layer (Figure 5.10b). The diameter of the hexagonal meshes is about 1 μm, and the width of the walls extends from 50 nm to 100 nm. The etching process was controlled by the use of Auger spectroscopy.

The transport properties of patterned GaAs exhibit interesting phenomena [30]. In a wide low-temperature regime up to 260 K, we find variable range hopping in two

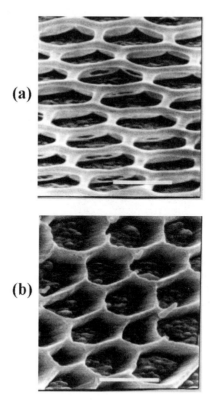

(a)

(b)

FIGURE 5.10. SEM images of (a) a nitrocellulose mask and (b) a GaAs network obtained by an ion-etching process. Scale bars: (a) 0.6 μm and (b) 0.75 μm.

dimensions with a nearly constant density of states of the defect band. The absolute values obtained by an adaptation of the local activation energy to predictions of theory are in good agreement with those reported in literature on thin bulk material. We have also analyzed the nonlinear voltage dependence of the differential conductivity by percolation theory. The onset of nonlinearity can be described by a threshold, i.e., a critical current. Between the local activation energy and the critical current, a good correlation of the different temperature regimes is found for which a transport mechanism is valid.

5.6. CONCLUSION

An experimental preparation technique was described that is capable to produce mesoscopic polymer network structures with different shapes of the basic cell: A drop of the initial polymer solution spreads onto a cooled water surface, and the water vapour interacts with the resulting polymer thin film. Following the self-organization process of precipitating droplets of the water vapour on the polymer layer, pulling the latter to the water droplets, and subsequently evaporating the solvent, the originally homogeneous polymer film proceeds to a hexagonal network pattern. It was demonstrated that the size of the basic hexagonal cell is determined by the diameter of the water vapour droplets used during preparation. It was suggested that the stabilization of water droplets on the fluid surface is indispensable for ordered structure formation. This is performed by the ability of the polymer to precipitate at the solution–water interface. The properties of different polymer solutions are discussed that can influence a growth rate, a size and a form of condensing water droplets, and their interaction between each other. By the help of an elementary model study on the self-organized structuring process in the liquid polymer films, it was succeeded in specifying and interpreting the morphology of the basic network cells observed experimentally.

ACKNOWLEDGMENTS

I acknowledge with much thanks my colleague I. Bashmakov for successful cooperation and fruitful discussions. I would like to thank G. H. Bauer and J. Parisi for very helpful discussions, and moreover to V. Uchov and S. Martyna for taking the scanning electron microscopy pictures.

REFERENCES

[1] R.R. Bhave, *Inorganic Membranes: Synthesis, Characteristics and Applications*, Van Nostrand Reinhold, New York, 1991.
[2] E. Yablonovitch, Inhibited spontaneous emission in solid-state physics and electronics, Phys. Rev. Lett. **58**(20), 2059–2062 (1987).
[3] N. Akoezbek and S. John, Optical solitary waves in two- and three-dimensional nonlinear photonic band-gap structures, Phys. Rev. E. **57**(2), 2287–2319 (1998).
[4] K. Busch and S. John, Photonic band gap formation in certain self-organizing systems, Phys. Rev. E **58**(3), 3896–3908 (1998).
[5] T. Bitzer, *Honeycomb Technology*, Chapman and Hall, London, 1997.
[6] S.A. Jenekhe and X.L. Chen, Self-assembly of ordered microporous material from rod–coil block copolymers, Science **283**, 372–375 (1999).
[7] Y. Xia, B. Gates, Y. Yin and Y. Lu, Monodispersed colloidal spheres: old materials with new applications, Adv. Mater. **12**(10), 693–713 (2000).
[8] D.J. Norris and Yu, A. Vlasov, Chemical approaches to three-dimensional semiconductor photonic crystals, Adv. Mater. **13**(6), 371–376 (2000).
[9] A. Steyer, P. Guenoun and D. Beysens, Hexatic and fat-fractal structures for water droplets condensing on oil, Phys. Rev. E. **48**(1), 428–431 (1993).
[10] G. Widawski, M. Rawiso and B. Francois, Self-organized honeycomb morphology of star-polymer polystyrene films, Nature **369**, 387–389 (1994).
[11] B. Francois, O. Pitois and J. Francois, Polymer films with a self-organized honeycomb morphology, Adv. Mater. **7**(12), 1041–1044 (1995).
[12] O. Pitois and B. Francois, Formation of ordered micro-porous membranes, Eur. Phys. J. B **8**, 225–231 (1999).
[13] O. Karthaus, N. Maruyama, X. Cieren, M. Shimomura, H. Hasegawa and T. Hashimoto, Water-assisted formation of micrometer-size honeycomb patterns of polymers, Langmuir **16**(15), 6071–6076 (2000).
[14] M. Srinivasarao, D. Collings, A. Philips and S. Patel, Three-dimensionally ordered array of air bubbles in a polymer film, Science **292**, 79–83 (2001).
[15] L.V. Govor, I.B. Butylina, I.A. Bashmakov, I.M. Grigorieva, V.K. Ksenevich and V.A. Samuilov, in *Advanced Semiconductor Devices and Microsystems*, Ed. T. Labinsky, Smolenice, Slovakia, 1996, pp. 81–83.

[16] L.V. Govor, I.A. Bashmakov, F.N. Kaputski, M. Pientka and J. Parisi, Self-organized formation of low-dimensional network structures starting from a nitrocellulose solution, Macromol. Chem. Phys. **201**(18), 2721–2728 (2000).

[17] L.V. Govor, I.A. Bashmakov, R. Kiebooms, V. Dyakonov and J. Parisi, Self-organized networks based on conjugated polymers, Adv. Mater. **13**(8), 588–591 (2001).

[18] A.W. Adamson, *Physical Chemistry of Surfaces*, Wiley, New York, 1982.

[19] D.Y.C. Chan, J.D. Henry and L.R. White, The interaction of colloidal particles collected at fluid interface, J. Colloid Interface Sci. **79**(2), 410–418 (1981).

[20] C.M. Knobler and D. Beysens, Growth of breath figures on fluid surfaces, Europhys. Lett. **6**(8), 707–712 (1988).

[21] H.M. Princen, in *Surface and Colloid Science*, Ed. E. Matijevic, Vol. 2, Wiley-Interscience, New York, 1969, pp. 1–84.

[22] D. Beysens and C.M. Knobler, Growth of breath figures, Phys. Rev. Lett. **57**(12), 1433–1436 (1986).

[23] F. Family and P. Meakin, Scaling of the droplet-size distribution in vapor-deposited thin films, Phys. Rev. Lett. **61**(4), 428–431 (1988).

[24] B.J. Briscoe and K.P. Galvin, The evolution of a 2D constrained growth system of droplets-breath figures, J. Phys. D: Appl. Phys. **23**(4), 422–428 (1990).

[25] A.V. Limaye, R.D. Narhe, A.M. Dhote and S.B. Ogale, Evidence for convective effects in breath figure formation on volatile fluid surfaces, Phys. Rev. Lett. **79**(20), 3762–3765 (1996).

[26] A. Steyer, P. Guenoun and D. Beysens, Two-dimensional ordering during droplet growth on a liquid surface, Phys. Rev. B **42**(1), 1086–1089 (1990).

[27] L.V. Govor, M. Goldbach, I.A. Bashmakov, I.B. Butylina and J. Parisi, Electrical properties of self-assembled carbon networks, Phys. Rev. B **62**(3), 2201–2208 (2000).

[28] L.V. Govor, I.A. Bashmakov, K. Boehme, M. Pientka and J. Parisi, Coulomb gap and variable-range hopping in self-organized carbon networks, J. Appl. Phys. **90**(3), 1307–1313 (2001).

[29] L.V. Govor, I.A. Bashmakov, K. Boehme and J. Parisi, Electrical field dependence of hopping conduction in self-organized carbon networks, J. Appl. Phys. **91**(2), 739–747 (2002).

[30] L.V. Govor, M. Goldbach, I.A. Bashmakov and J. Parisi, Preparation and electrical characterization of low-dimensional net structures made out of GaAs epitaxial layers, Phys. Lett. A **261**, 197–204 (1999).

6

From Nanosize Silica Spheres to Three-Dimensional Colloidal Crystals

Siegmund Greulich-Weber and Heinrich Marsmann
Department of Physics, Faculty of Science, University of Paderborn, D-33095 Paderborn, Germany
greulich-weber@physik.upb.de

6.1. INTRODUCTION

Nanometre-scale periodic porous structures exhibit many unique optical, electrical and mechanical properties that can be exploited in a wide range of applications from photonics and electronics to biological and medical sensing. The synthesis of sub-micrometre building particles and their assemblies, such as silica or polystyrene spheres [1–8], nanotubes [9–15], nanowires [16–22] and nanocrystals [23–26], are of significant importance for the development of advanced nanotechnology. It is worthwhile to notice that there still remains a great challenge in synthesizing nanoporous materials with highly and precisely controlled pore sizes in ordered three-dimensional structures in large volumes [27–31]. A promising technique for fabricating 3D nanoporous structures is based on the self-assembling growth of monodisperse spherical colloidal particles of silica or polystyrene, further on called nanospheres. From such three-dimensional periodical packages, so-called colloidal crystals or opals, useful nanostructured materials can be created by replicating these crystals in a durable matrix that preserves their key feature of a long-range periodic structure. The fidelity of this procedure is mainly determined by the crystal growth mechanisms of opals and the monodisperity of nanospheres. Therefore in Section 6.2 methods for obtaining and modifying monodisperse silica spheres are reviewed [32–35], which mainly rely on the process invented by Stoeber [6]. Latex spheres are usually obtained by emulsion polymerization, which was described in detail elsewhere [36–42]. In Section 6.3 important mechanisms for growing opals from nanospheres are discussed. Since the dense packing (fcc, hcp) of monodisperse spheres creates regular voids, these crystals already belong to the family of nanoporous materials. By using colloidal crystals as templates inverted structures can be created by filling the voids with other materials and afterwards releasing the nanospheres. The porous materials obtained by this approach have also been referred to as "inverted opals" or "inverse opals" because they have an open, periodic 3D framework complementary to that of an opal structure. Their fabrication, properties and applications are discussed in Section 6.4. Finally, in Section 6.5 a by far not complete review of actual and promising applications is given.

6.2. SYNTHESIS OF COLLOIDAL SILICA NANOSPHERES

Many compounds of the 6th group of the periodic table of the elements such as oxides, sulphides, etc. are insoluble in aqueous solutions. This behaviour can be used to generate small particles by chemical reactions. The properties of a number of metal oxide particles have been described in [1–3,43]. Depending on their crystallinity, spheres, cubes or needles of several nanometres diameter are observed. Because of quantum size effects the small particles build up have different properties compared to bulk material. If they are small enough they stay in solution forming sols. Prominent metal colloids are build, e.g., from gold, which displays red colour. Spheres of diameters up to several hundred nanometres are grown from silicon dioxide. The synthesis and properties of these amorphous silica spheres are the topic of large number of review articles [4,5]. Especially the observation made by Stoeber *et al.* [6] that silica spheres can be grown almost perfectly as monodisperse colloidal spheres has fascinated meanwhile generations of researchers. The synthesis and modification of these nanometre-sized silica spheres will be discussed in the following.

6.2.1. Synthesis

Generally the formation of silica is a two-step process starting from a source of silicic acid and followed by a condensation resulting in an amorphous gel particle. This condensation process leads to individual particles or a gel spanning the whole container. Most of the syntheses start from an organic ester of the silicic acid, e.g., tetraethoxisilane, $(C_2H_5O)_4Si$, usually abbreviated to TEOS. The first step is the hydrolysis, which is followed by the condensation over the silanol groups:

$$\tag{6.1}$$

$$\begin{array}{c}
\text{O-Si-OH + HO-Si-O} \longrightarrow \text{O-Si-O-Si-O} + H_2O
\end{array}$$

(6.2)

Because TEOS has four ethoxy groups three- and fourfold reactions lead to 3D networks. Both reactions are moderated by a catalyst, usually ammonia, NH_3, as in the original Stoeber approach [6] or acetic acid as used in [7] or hydrochloric acid as described in [44]. Further sources suitable as silicic acid are feasible, e.g., aqueous solutions of water glass, or other esters of silicic acid, e.g., tetramethoxisilane, $(CH_3O)_4$, TMOS (because of its highly poisonous nature the use of TMOS is not encouraged). The overall reaction can be dissected into two parts. The first is the initiation step where the primary particles are formed which is followed by the growth of the particles as long as silicic acid is available.

The formation of monodisperse, uniform and smooth spheres in the Stoeber process has attracted much interest. The reactions take place in a mixture of water and a lower alcohol such as ethanol with a varying amount of ammonia. Depending on the concentration of the components of the reaction and the reaction temperature spheres with diameters between 50 and 2000 nm have been synthesized. It was observed that the smallest particles grow best in methanol, while the largest in butanol. An increase of the concentration of ammonia caused larger diameters of the spheres, while the amount of water was of minor influence [6].

Basically there are two theories explaining the particle growth. One involves the accumulation of monomeric or small condensates of silicic acid to form primary particles. It is assumed that during growth no further primary particles are formed and that the growth of the particles occurs by aggregation of monomers [45–48]. The particle growth is reaction limited and depends on the reaction rate of the hydrolysis of the silicic acid ester.

According to an alternative theory, there is a continuous generation of primary particles which merge with larger particles causing their further growth [49,50]. A study of the particle growth by cryogenic transmission electron microscopy clearly shows that small, low-density particles coalesce forming primary particles. However, in the following, no evidence of smaller, dense particles besides the large growing spheres was observed [51]. Possible formation of irregular-shaped particles was claimed to be the result of an aggregation of larger particles [48]. The simultaneous growth of silica particles of different sizes is possible if the concentration of the alkoxide is kept low. If no new particles are generated, all particles grow with the same rate pointing to a surface-limited reaction. If the concentration of TEOS is high enough allowing the generation of new particles, the smaller particles will grow faster than the larger ones, thus indicating that their growth is diffusion limited [52]. A mathematical model describing the generation of monodisperse spheres explains the monodisperity as a function of the interfacial energy of the solution–particulate system without the need of a separate nucleation phase [53].

A precise control of the number of primary particles during the nucleation phase is difficult. Therefore the size of particles synthesized under identical conditions varies by about 20% in each batch, although they are still monodisperse. The most successful procedure of obtaining a certain size of spheres is to start from a solution of spheres smaller than intended and use them as a seed. The gradual addition of TEOS using Stoeber conditions lets the spheres grow to the desired size [48,55].

Besides the usual procedure synthesizing silica particles by hydrolysis or condensation in alcohol water mixtures, silica spheres may also be grown in micelles. The reaction medium consists of water and, e.g., an ionic surfactant such as benzethonium chloride [56], a non-ionic one such as nonyl phenol ethylene oxide [46,57] or a block copolymer and several alcohols [58]. The advantage of such a procedure is the possibility of obtaining nanospheres of difficult monomers, e.g., a mixture of TEOS and methyltrimethoxysilane. The size of the particles is determined by the ratio of surfactant to monomer [46,56,57] or the flexibility of the organic part of the alcohol [58]. It seems that very small particles can be obtained with the help of micelles. Diameter of about 10 nm has been reported [56]. It is remarkable that cetyltrimethylammonium bromide as a surfactant yields with TEOS and water irregular mesoporous silica spheres with highly ordered pores while adding ethanol to the reaction mixture gave smooth spheres with starburst pores [59]. Silica spheres with zeolite-type mesopores are also synthesized with the help of surfactants [60,61].

^{29}Si magic angle nuclear magnetic resonance (MAS-NMR) affords insight into the microstucture of the silica spheres. Because of the fact that silicon connected by four siloxane bonds to other silicon atoms (Q^4 groups) has a chemical shift (~ -110 ppm) different to silicon connected to three (Q^3 groups, ~ -100 ppm) or two silicon atoms (Q^2 groups, ~ -90 ppm). It was observed that the silica prepared according the Stoeber process contains about two thirds of Q^4 groups and one third of Q^3 groups and only a few per cent of Q^2 groups. Silica obtained with the help of nonyl phenol ethylene oxide has a more open structure, due to the higher content of Q^3 groups. The open valences of Q^3 and Q^2 groups are saturated by ethoxy residues confirmed by ^{13}C MAS-NMR [4].

6.2.2. Modification of Silica Spheres

As interactions between silica spheres are required to tune their physical properties according to the needs of intended applications, it is necessary to design their chemical composition. Furthermore, the spheres have to be dried for almost all applications, which might change their properties drastically. The structure, size and composition of these hybrid particles can be altered in a controllable way to tailor their optical, electrical, thermal, mechanical, electro-optical, magnetic and catalytic properties over a broad range. In the following several ways to achieve appropriate requirements are pointed out.

6.2.2.1. Modification with Organic Residues.

The motivation to modify silica particles with organic residue is mainly due to the following reasons.

The reaction between two silanol groups (Equations (6.1) and (6.2)) might also happen when two silica particles touch each other. This leads to irreversible connections between the particles and will prevent resuspension in a solvent. By modifying the surface with organic groups the surface of the particles will be protected and the particles may receive new properties suitable for the surrounding solvent.

An organic modification might be realized in order to obtain new functionalities, e.g., for bonding with polymers or with chromophores.

One simple example is the surface modification of silica spheres with stearyl alcohol, a long-chain alcohol [62,63]. Also octadecanol was used to form a protective shell around the silica particles [60], and methods were used to attach polystyrene chains via silanol groups [64] or polyisobutene [65]. In that way silica particles were made more compatible to organic solvents and their surfaces deactivated. More versatile are procedures of introducing organic functional groups. Rather easy to synthesize and commercially available are trialkoxyorganosilanes, where alkoxy groups can react with the silanol groups of silica, forming surface modified silica. Alternatively they might be mixed with the TEOS reacting to a copolymer. The organic chain mostly a propyl residue carries a functional group R1 such as a halogen, an ammino or a mercapto group.

$$\begin{array}{c} \diagdown O \\ | \\ \diagup O-\underset{|}{\overset{}{\text{Si}}}\diagup\diagup\diagup^{R1} \\ O \\ \diagdown \end{array}$$

(6.3)

The modifiers and examples given above or in the literature are tabulated in Table 6.1. Functional groups have been used to attach dyes such as fluorescein [60,66,67] or rhodamine [60] or to form a network in epoxy composite material [68]. It was suggested that the dye molecules (fluorescein) tend to form clusters during the synthesis of the colloidal spheres [66]. Reactions with trimethoxysilylated polymers such as poly(maleic anhydride-costyrene) or poly(methacrylate) lead to silica spheres with a coat of polymer.

There are interesting proposals to incorporate inorganic ions and compounds in the silica to make use of their optical properties. CuO was infiltrated in MCM-41-type silica nanospheres in a Stoeber-type synthesis of silica particles [61]. The chemical degradation of alkyldithiocarbamato complexes of cadmium in the presence of sub-micrometric SiO_2 particles results in the formation of CdS nanoparticles on the sphere surface [69]. Because most metal ions form precipitates in basic solution, the synthesis of silica particles under acidic catalysis is of advantage. The incorporation of rare earth ions in silica spheres is interesting for photonic purposes succeeded in that way. Pr^{3+} and Eu^{3+} could be found in silica spheres, when TEOS was reacted in a mixture of water and acetic acid. The

TABLE 6.1. Selected organic modifiers for silica.

Reagent	Reference
$NH_2(CH_2)_3Si(OC_2H_5)_3$	[28,29,30,33,34]
$NH_2(CH_2)_3Si(CH_3)(OC_2H_5)_2$	[32]
$HS(CH_2)_3Si(OC_2H_5)_3$	[28]
$ClC_6H_4Si(OCH_3)_3$	[31]
$CH_3C_6H_3(NCO)_2$	[32]
Epichlorhydrin	[32]

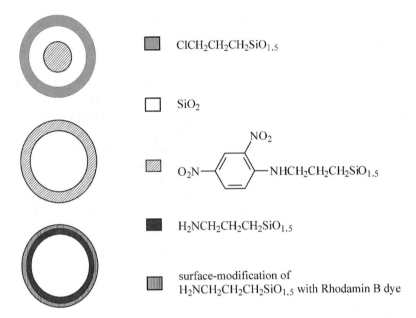

FIGURE 6.1. Schematic representations of different core-shell designs for silica spheres. Details are given in the text and in Table 6.2.

particles exhibited bright room-temperature luminescences when excited by an argon ion laser at 488 nm [70]. Er^{3+} and Tb^{3+} ions could be incorporated by an analogous procedure [71], while silver nanoparticles ($\sim$2–5 nm) inside silica spheres could be generated by photochemical reduction of silver nitrate [72].

6.2.2.2. Core-Shell Designs. The growth of silica spheres by a slow generation of silicic acid by, e.g., hydrolysis of TEOS could be used to form additional layers on them. Several design models are possible. An example is given in Figure 6.1. Several examples are possible and worked out. The interested reader is referred to the literature [73]. A recent review on nanoengineering of spherical particle surfaces was, e.g., given by Caruso [74]. Further special designs for various applications are given in [28,54,75–79]. A collection of selected layer structures is summarized in Table 6.2. Some interesting materials for photonic applications could be made this way. Especially promising seems the inclusion of highly refractive compounds such as metals, oxides and sulphides.

    The core-shell design is a clever tool to modify the properties of the particles. For instance, a chromophore can be put on the shell to interact with chemical influences from outside while by placing the same modifier inside it will shield against the environment. It is possible to tailor a composite particle in which every layer will confer a special property to it [33,38].

6.2.2.3. Infiltration and Doping. Besides the core-shell method, which allows a modification of silica spheres during the sphere synthesis, other materials such as semiconductor nanocrystallites, metal sols and organic chromophores have been incorporated

TABLE 6.2. Examples of core-shell silica particles.

Core	1th layer	2nd layer	Reference
SiO_2 + APS[a]	SiO_2		[30]
SiO_2 + APS + Dye	SiO_2		[33]
SiO_2	SiO_2 + APS + Dye		[33,38]
SiO_2	SiO_2 + APS + Dye	SiO_2 + octadecanol	[33,38]
SiO_2	Ag		[21]
$\alpha - Fe_2O_3$	SiO_2		[43]
ZnS	SiO_2		[44]
SiO_2	ZnS		[44,45]
SiO_2	SiO_2 + CdS	SiO_2	[39]

[a]3-Aminopropyltriethoxysilane.

into silica colloids during the synthetic step in order to functionalize these particles with useful optical, electrical or magnetic properties [75–77]. However, the distribution of the materials incorporated is less homogeneous in comparison to the core-shell method.

Another way of modifying nanospheres usually used for catalytic applications is to infiltrate them with suitable materials after their synthesis [80–87]. This can be done by either adding salts to the solvent containing the material intended to deposit in the porous nanospheres or by allowing a chemical reaction of cations with the inner surface of the spheres or the residual organic content left from the sphere synthesis. The latter is present even after careful cleaning of the spheres after their synthesis. However, if the desired material is incorporated into the sphere's silica matrix, one may call this doping of silica. Most probably this will not happen at room temperature. Heating the material will allow the doping ions to further diffuse into the porous silica (see the next section). Such a kind of doping is considerably more stable than "fixing" the infiltrated material at some residual organic content which might partly evaporated during drying the spheres even at room temperature.

With the further on so-called galvanic method electric-field-driven ions in the solvent are captured by the porous spheres between two electrodes [88,89]. The galvanic method works most efficiently for metal ions and allows an effective infiltration at room temperature allowing afterwards the same chemical reactions as described above or additional temperature treatments.

There are a few more expensive methods to modify the properties of silica spheres such as, e.g., ion implantation [90–93]. Another less expensive method that finally should be mentioned is infiltration of dried spheres in a dopant gas phase, which of course is most effective at high temperatures [87].

6.2.2.4. Thermal Treatment. The spheres in an opal structure as discussed in Section 6.3 are held together just by relatively weak forces. For applications cracking or disrupting the structure is always a problem. By sintering, annealing or heating silica spheres above the glass point it is possible to strengthen the material. Pristine samples of silica colloids will undergo a series of changes when they are thermally treated at elevated temperatures. First the absorbed water will be released at ~150 °C. In the temperature range of 400–700 °C the silanol groups will be crosslinked via dehydration. Finally,

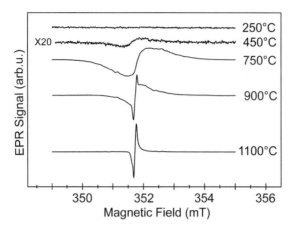

FIGURE 6.2. EPR spectra of undoped silica spheres (1200 nm diameter) measured at $T = 10$ K and a microwave frequency of 9.42 GHz [87]. As grown silica spheres do not show an EPR spectrum (upper trace). After annealing at different temperatures for 1 hour each new EPR spectra appear, which are discussed in detail in the text.

these particles will start to fuse into aggregates when the temperature is raised above the glass transition temperature of amorphous silica ($\sim$800 °C). These steps are clearly observable with suitable spectroscopic methods. Electron paramagnetic resonance (EPR) turned out to be a particularly helpful tool, showing all steps as presented in Figure 6.2 [87]. While after drying the spheres no EPR was observed, around 400 °C an EPR signal appeared, which again decreased at the expense of a completely different signal which became strongest above 1000 °C. The broad and asymmetric signal observed at around 750 °C (Figure 6.2) is a typical signal of an amorphous surface defect in the porous silica matrix [87]. The EPR spectrum shown for annealing at 1100 °C (see Figure 6.2) has a comparably small line width and is symmetric with respect to the baseline, being more typical for a crystalline or polycrystalline environment. Further small satellite lines, not shown in Figure 6.2, are due to the interaction of an unpaired electron with two different ^{29}Si neighbours. A similar spectrum is known from the so-called E′ centre in SiO$_2$ [94], a dangling bond centre at the SiO$_2$ surface. Figure 6.2 implies that the EPR may serve as a tool for controlling the heat treatment of silica spheres, which is unavoidable for the realization of nanoporous materials discussed in the next sections.

The transition from a less dense amorphous gel composition to a glass-like structure is also seen in the optical transmission spectra, showing that the spheres became more transparent in the visible range after annealing at 1000 °C (Figure 6.3) [87].

Finally the transition from the dried gel state to the glass-like state results in a further decrease in sphere diameter ($\sim$5%), which was demonstrated by electron microscopy as shown in Figure 6.4 [87].

Spectroscopic control becomes even more important in the case of sphere infiltration or doping. As an example Figure 6.5 shows EPR spectra of Cu-infiltrated silica spheres. The lower trace in Figure 6.5 is observed without heat treatment either after Cu-infiltration using copper chloride, copper sulphate or copper nitrate as solvents or after infiltration by the galvanic method [87]. The latter turned out to be the most efficient one. The EPR

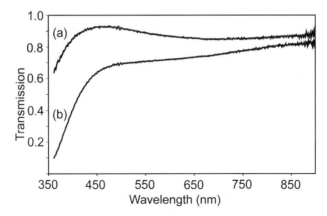

FIGURE 6.3. Transmission spectra of undoped dried silica spheres (400 nm diameter) otherwise as grown (a) and after annealing for 1 hour at 1000 °C (b) [87].

spectrum (Figure 6.5, lower trace) belongs to a Cu^{2+} complex containing nitrogen [95] in an amorphous environment (strictly speaking hexamine copper nitride). Independent of the infiltration process used the EPR spectrum completely vanished after heat treatment at about 1000 °C. Along with that the colour of the silica spheres changed from light blue to dark red. A similar change in colour is observed if untreated spheres are electron-irradiated; however, in that case a new EPR spectrum is observed probably due to an intrinsic defect [87].

Though we end up with the same colour, obvious different reasons accidentally lead to the same colour. Optical spectroscopy helps showing different transmission spectra. As

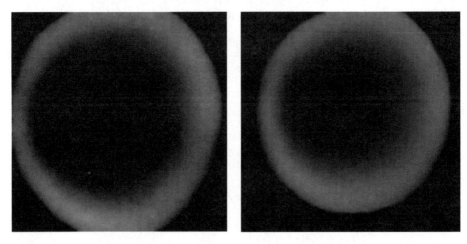

FIGURE 6.4. Transmission electron microscopy (TEM) images of single spheres which as grown had a diameter of 1.2 μm (as determined by dynamical light scattering, Huber). The left image shows a sphere after drying. The total size of the image is $1 \times 1\,\mu m^2$. The right image shows one sphere from the same batch of silica spheres but after heat treatment at 1000 °C for 1 hour [87].

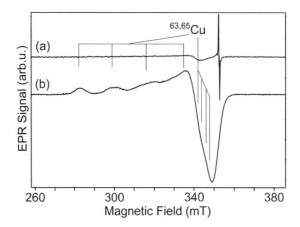

FIGURE 6.5. EPR spectra of Cu-infiltrated silica spheres (for details see the text), measured at T = 10 K and a microwave frequency of 9.42 GHz. As grown and afterwards dried independent of the fabrication procedure a Cu-related EPR spectrum is observed (lower trace). After heat treatment at T > 400 °C a new EPR spectrum appears. Without annealing, but with electron irradiation the same EPR spectrum is observed. Details are discussed in the text. Further information on the parameters describing the EPR is given in [87].

an example Figure 6.6 presents the spectrum obtained after Cu-infiltration and a following heat treatment, showing a plasmon resonance at around 600 nm. This indicates that the Cu^{2+} EPR signal vanished because metal colloids are built due to the heat treatment [87]. The red colour appearing after electron irradiation is a result of an indirect charge transfer effect.

Cu-infiltrated silica spheres illustrate the need to control the modification of spheres by adequate methods. The most helpful tool for controlling the actual realization of sphere modifications is EPR resolving the microscopic and electronic structure of the dopants and their local environment with high sensitivity. Complementary optical investigations

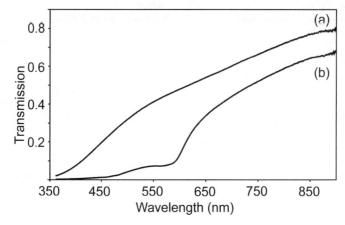

FIGURE 6.6. Transmission spectra of Cu-doped and dried silica spheres (400 nm diameter) (a) and after annealing for 1 hour at 1000 °C (b) [87].

are indispensable. Furthermore, geometrical variations are best observed via electron microscopy; under best conditions also details of the dopant distribution are observed.

So far using preparation methods as described above several materials have been successfully incorporated into silica spheres for different applications. Besides the use for medical [95–97] or catalytic applications [61,98] there are promising developments based on magnetic particles [99–104] or metal ions [99,101,104–108]. For most applications aggregated spheres, often periodically arranged, are needed instead of single spheres. Thus capabilities to fabricate ordered structures of spherical colloids are discussed in the following section.

6.3. GROWTH OF COLLOIDAL CRYSTALS

In nature silica spheres sometimes crystallize into colloidal crystals of fcc structure through self-assembly known as opals. In the following we describe procedures for growing artificial opals [109–113] and so-called inverted opals [27,114–124], which are negative replica of opal structures. In principle, there are three common techniques for obtaining artificial opals from monodisperse silica nanospheres stored in protic solvents. Since the density of SiO_2 is significantly larger than that of water (or related hydrogen solvents) sedimentation is the dominating method for growing bulk crystals [125]. On the other hand, the silica spheres experience attractive capillary forces at the solvent surface due to evaporation of the solvent, leading to self-assembly of spheres close to the surface. This method is commonly used for growing thin film opals. Both methods critically depend on competing interactions between spheres themselves and between spheres and environment. Another attempt is to ignore these interactions by applying strong external forces such as centrifugation [118,126,127] or electrophoresis [128–130]. However, in this case no self-assembly of the spheres is expected and additional agitation is needed to improve crystalline quality.

Opal growth by self-organization of monodisperse silica spheres is of course the most interesting and by far most intensive studied phenomenon. Charged silica colloids suspended in a protic solvent interact through hard-core repulsions, van der Waals attractions, Coulomb interactions and hydrodynamic coupling. The influence of the spheres on the surrounding medium modifies these interactions, for instance leading to screening of Coulomb interactions or giving rise to entropically driven depletion interactions in heterogeneous suspensions. Solvents usually used are, e.g., water or ethanol.

In the following relevant interactions responsible for growth mechanisms of colloidal crystals from a diluted suspension of monodisperse silica spheres will be represented briefly.

6.3.1. Interactions Between Colloidal Particles

6.3.1.1. Coulomb Interaction. The acidity of the silanol groups causes a negative charge of the silica particles [5]. The isoelectric point of the particles can be altered by surface modification, e.g., for a pH value of about 3 for amino group functionalized particles [4]. Therefore silica spheres always carry surface charges due to adsorbed ions or electrons, and due to the dissociation of polar groups at the surface the Coulomb interaction is the most important. Furthermore, the solvent itself contains moveable charges (dissolved salts). Unaffected by the kind of charges dissolved ions are mainly found at the complementary charged silica spheres. A shell of ions is precipitated, the so-called Stern layer [131]. The strength of this Coulomb energy dissipates exponentially away from the surface as the concentration of further ions of this layer follows Boltzmann's law. Using this space charge distribution the resulting potential can be derived by the Poisson equation. The mathematics that describes the electrostatic force finds its origin in the Poisson–Boltzmann distribution [132]:

$$\nabla^2 \phi(r) = \kappa^2 \phi(r). \tag{6.4}$$

Numerous models have been developed from this fundamental relationship. Simplified models are based on assumptions about particle geometry, surface charge and potential [133]. An electrostatic interaction energy model used to describe the commonly encountered geometry of a sphere and a flat plate in water is, for example, given by Gregory [134] (see also [133]).

Solving the Poisson–Boltzmann equation for the long-range Coulomb repulsion shielded by electrolytes outside a sphere of radius R carrying charge $z_1 e$ yields a Yukawa potential [135]:

$$\phi(r) = \frac{z_1 e}{4\pi\varepsilon(1+\kappa R)} \frac{\exp[-\kappa(r-R)]}{r}. \tag{6.5}$$

The potential is repulsive and steep compared to the sphere radius. The particle charge can be determined easily from the mobility in an electric field and is about $10^{-16}\,\text{C} = 10^3 e^-$. The remaining ions in an otherwise ion-poor solvent do not influence the estimation made above [136].

κ in Equations (6.4) and (6.5) is the reciprocal Debye–Hückel length of the solvent [137]

$$\kappa^2 = \frac{e^2}{\varepsilon k_{\mathrm{B}}T} \sum_{i=2}^{\infty} n_i z_i^2 \tag{6.6}$$

with z_i the valence and n_i the concentration of ion i. An increase in electrolyte concentration results in a decrease in the Debye–Hückel length and concomitant reduction in the electrostatic interaction energy due to the shielding effect of the ion shell of the sphere on the Coulomb potential. At a specific salt concentration, known as the critical coagulation concentration, the electrostatic interaction force can be virtually neutralized. Changes in pH influence the range and magnitude of electrostatic forces. Reactions between protons and charged surface functional groups can change the net surface potential on a particle [138].

6.3.1.2. Van der Waals Attraction.

The van der Waals attraction is a common interaction of all matter. In the case of silica spheres one has to integrate over all dipoles of spheres, which can be done easily because of the simple geometrical properties of a pair of spheres. The interaction energy results in the following attractive potential [139],

$$\phi(r) = -A_{12}\frac{R}{12r} \tag{6.7}$$

with the sphere radius R, the distance r between two spheres and the constant A_{ij} for the material combination. A_{ij} is known as the Hamaker constant [140,141] and is proportional to the square of the polarization of a material. In the theory of Hamaker the constant A is given as the sum of A_{ij} over single terms of the combination of two materials (ij) which are tabled [142]. For two particles consisting of materials 1 and 2, respectively, in a solvent (medium 3) the Hamaker constant is calculated by [132,143]

$$A_{123} = A_{12} + A_{33} - A_{13} - A_3. \qquad (6.8)$$

So far the competing electrostatic interactions, the screened Coulomb and the attractive van der Waals interaction (now expanded including dispersion effects) have been discussed as the main interactions between charged objects in an electrolyte. However, further interactions have to be considered (see below) while it was recently argued that both electrostatic interactions discussed so far cannot be considered separately [144–146].

Attempts to quantitatively describe colloidal interactions [147,148] resulted in the famous DLVO (Derjaguin–Landau–Verwey–Overbeek) model. The DLVO theory was developed by balancing attractive dispersion (London, van der Waals) and repulsive electrostatic Coulombic forces and meanwhile had been accepted to be inclusive of all primary interfacial forces of significance. Both van der Waals and Coulombic forces can be either repulsive or attractive depending on chemical structure, suspending medium properties and surface potential.

6.3.1.3. DLVO Theory. The nonlinear Poisson–Boltzmann equation (6.4) has only been solved analytically for a very restricted set of geometries, parallel charged plates, for example [133,134]. Solutions for more general geometries such as pairs of spheres have proved elusive. Even this intractable model involves a dramatic simplifying approximation. The suspending fluid appears in Equations (6.4) and (6.5) only through its dielectric constant. This so-called primitive model completely neglects effects due to the structure of the solvent, an approximation which fails when the separation between nanospheres becomes comparable to a few molecular radii.

Derjaguin, Landau, Verwey and Overbeek (DLVO) [147,148] pushed the field forward in the 1940s by applying approximations from the Debye–Hückel theory of electrolyte structure [137,139].

The DLVO theory provides approximate solutions to the Poisson–Boltzmann equation describing the nonlinear coupling between the electrostatic potential and the distribution of ions in a colloidal suspension. It predicts that the effective pair interaction in dense suspensions sometimes has a long-ranged attractive component. The resulting interaction between isolated pairs of well-separated spheres has the form [149]

$$\frac{\phi(r)}{k_B T} = Z_1^* Z_2^* \frac{e^{\kappa a1}}{1 + \kappa a_1} \frac{e^{\kappa a2}}{1 + \kappa a_2} \lambda_B \frac{e^{-\kappa r}}{r} + \frac{V(r)}{k_B T} \qquad (6.9)$$

where r is the centre-to-centre separation between two spheres of radii a_i with effective surface charges Z_i^* in an electrolyte of the Debye–Hückel screening length κ^{-1}. $\lambda_B = e^2/k_B T$ is the Bjerrum length, equal to 0.714 nm in water at $T = 24\,°C$, and $V(r)$ is the Coulomb potential.

The full DLVO potential includes a term accounting for dispersion interactions (see above, extended van der Waals interaction), which, however, is negligibly weak for well-separated spheres [150,151].

6.3.1.4. Depletion Potential.   Besides optimizing the above-described interactions between colloidal particles themselves and the solvent, further procedures are necessary for the growth of extended colloidal crystals of high quality. It was shown by Asakura and Oosawa [152] and independently recovered and further elaborated by Vrij [153] that the addition of a non-adsorbing polymer to a dispersion of colloidal particles will lead to an effective attractive interaction. The attractive interaction, which is called the depletion interaction, is the origin of a rich phase behaviour displayed by colloid–polymer mixtures. Depletion interactions can occur in systems that have particles with disparate sizes: for example, a system that contains large spheres in dispersion comprised relatively small colloids. As the large spheres approach one another, the smaller colloids will be excluded from the gap between them, which results in a decrease in osmotic pressure between the spheres. This reduction in osmotic pressure results in an attractive force called the depletion attraction. In Figure 6.7 this situation is sketched schematically.

The depletion interaction can be explained in terms of purely repulsive interactions between the polymers and the much larger colloidal particles. Each colloidal particle is surrounded by a shell with a thickness of the order of the radius of gyration of a polymer molecule through which the centre of the polymer cannot penetrate. This excluded region is called the depletion zone (see Figure 6.7a). When two colloidal particles approach each other such that their respective depletion zones start to overlap, the available volume for the polymer increases. This extra volume in turn causes the total entropy to increase and the free energy to decrease. The colloidal particles experience an effective attraction. In other words, if the depletion layers overlap, the osmotic pressure becomes anisotropic and there is a net osmotic force (see Figure 6.7b).

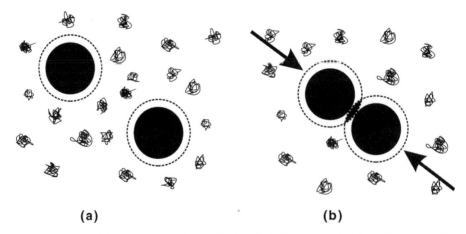

(a)   **(b)**

FIGURE 6.7. Schematic representation of two colloidal spheres in a polymer solution with non-adsorbing polymers. The depletion layers are indicated by dashed circles. (a) The osmotic pressure on the spheres due to the polymer solution is isotropic when there is no overlap. (b) The osmotic pressure on the spheres is unbalanced for overlapping depletion layers (the excess pressure is indicated by the arrows).

In the case of binary mixtures of large and small colloidal hard spheres, the depletion potential is in lowest order in the density given by [154,155]

$$\phi(r) = 3k_B T \delta \frac{R}{\sigma} \left(1 - \frac{r}{\sigma}\right)^2. \tag{6.10}$$

Here, $k_B T$ is the thermal energy, δ is the volume fraction of the small spheres, R is the radius of the large spheres, σ is the diameter of the small spheres and r is the distance between the surfaces of the two large spheres.

The addition of non-adsorbing polymer to a colloidal suspension induces an interparticle "depletion" attraction whose range and depth can be "tuned" independently by altering the polymer's molecular weight and concentration respectively and could therefore be an extremely helpful tool for the growth of colloidal crystals. The control of attraction due to the density of much smaller particles leads to large crystalline areas in separated phases [156]. In the case of polymer particle several investigations on the phase behaviour were made [157–159] showing useful phase diagrams [159] for crystal growth. A consequence from these phase diagrams is that already at low bead densities the crystallization is initialized when a high concentration of small particles is present in the solvent.

6.3.2. Dynamics in Colloidal Suspensions

6.3.2.1. Gravity and Friction. Optimal fabrication of useful devices often requires an understanding of how to control the crystalline structure by changing the interactions among colloidal spheres, as well as the kinetics of the colloidal spheres. Particles in a suspension are always moving independently due to the Brownian motion. Einstein [156] showed that colloidal particles in a stationary solvent can be regarded as an ideal gas, which results in a diffusion coefficient of a highly diluted system under the assumption that motion through the solvent is connected with the Stokes friction.

However, the suspension additionally experiences gravity. Altogether one has to consider the interaction of gravitational ($F_g = (1/6\pi)\rho_s g d^3$), Archimedes ($F_A = (1/6\pi)\rho_h g d^3$) and frictional forces ($F_f = 3\pi \eta v d$), where ρ_s and ρ_h are the sphere and solvent mass densities, g is the gravity acceleration, η is the viscosity of the solvent, d is the sphere diameter and v is their velocity. When all forces are balanced, the Stokes law is obtained giving directly an expression for the velocity of the silica spheres in the solvent:

$$v = \frac{g d^s}{2\eta}(\rho_h - \rho_s). \tag{6.11}$$

Experimental observations reproduced this expression in an excellent way [160]. Best colloid crystal qualities are obtained for sedimentation speeds between 0.2 and 0.7 mm/h [130]. This can be achieved by appropriate chosen ratio of densities of the silica spheres and the solvent partly compensating the gravitational field. If the suspension is sufficiently diluted the previously discussed attractive and repulsive forces between spheres result in crystallized structures of high order. Especially the formation of the first sedimentation layer was documented in detail by Salvarezza *et al.* [161]. Even at the surface the spheres move by diffusion until they reach their final position. From diffusion theory a distribution probability of spheres can be calculated which then can be used to study the crystallization of the system. Similar to atomic crystallization the ratio of the total volume and that of all particles in the solvent can be used as a thermodynamic parameter of state in order to describe the phase transition from liquid to solid [162,163]. If this volume ratio exceeds the critical value of 0.49 it will become entropically favourable to establish a long-range order and crystallization starts [164–166]. The crystal structures that have been observed in such systems include bcc, fcc, rhcp and AB_2 [167–177]. When the screening length is shorter than the centre-to-centre distance between two spheres, the colloidal spheres act like "hard spheres" and they will not influence each other until they are in physical contact. In this case, an fcc crystal structure is formed. If the screening length is longer than the centre-to-centre distance, the colloidal spheres behave like "soft" spheres and will only crystallize into a bcc lattice [174–177]. In both crystalline structures, the colloidal spheres are separated by a distance comparable to their size. The disorder-to-order transition can be provoked either by increasing the volume fraction of colloidal spheres or by extending the range of the screening length.

The major disadvantage of the sedimentation method is that it has very little control over the morphology of the top surface and the number of layers of the 3D crystalline arrays [161]. It also takes relatively long periods of time (weeks to months) to completely settle sub-micrometre-sized particles. The 3D arrays produced by the sedimentation method are often polycrystalline. Sedimentation under an oscillatory shear could greatly enhance the crystallinity and ordering in the resulting 3D arrays [178]. In addition to its function as a means for increasing the concentration of colloidal spheres, the gravitational field has a profound effect on the crystalline structure. In a micro-gravity experiment this effect could be eliminated showing that the colloidal spheres were organized into a purely random hexagonal close-packed (rhcp) structure [179].

Figure 6.8 shows a part of a colloidal crystal made from monodisperse silica spheres by sedimentation deposited on a silica glass substrate. The displayed area proves that an

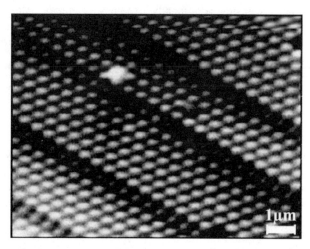

FIGURE 6.8. Electron micrograph of a colloidal crystal made of silica spheres (400 nm diameter) showing an fcc structure. The crystal was grown by sedimentation on a silica glass substrate. The crystalline quality could be improved by applying an external acoustic field [8].

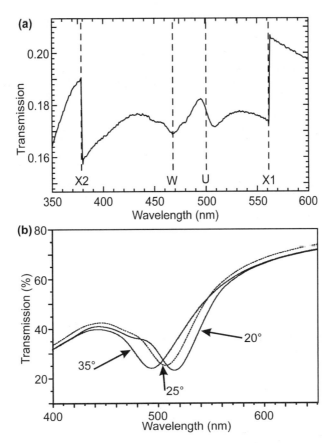

FIGURE 6.9. (a) Transmission spectrum of a polycrystalline colloidal crystal from silica spheres (240 nm diameter) [180]. The sizes of the crystals are at least 20 μm. Since the lattice constants of the colloidal crystals are in the range of the wavelength of the visible light, the transmission spectrum is mainly given by Bragg reflections. Therefore from the transmission spectrum the quality of a colloidal crystal can easily be determined without any damage. The spectrum can be calculated as a superposition of reflections from various differently oriented microcrystals [226]. The notation is given according to directions in k space for an fcc crystal structure. (b) Transmission spectrum of a highly ordered colloidal crystal of silica spheres (230 nm diameter) [8]. The angles are given relative to a $\langle 111 \rangle$ direction.

fcc crystal structure is realized. The high crystalline quality was, however, received by applying an external acoustic field [8]. The crystal quality can be controlled with high spatial resolution by electron microscopy as demonstrated in Figure 6.8, which unfortunately is a destructive method. Adapting X-ray diffraction used for the study of crystalline structures of atomic crystals to sub-micrometre lattice elements (silica spheres) visible or ultraviolet light has to be used. One receives similar diffraction images for colloidal crystals as one obtains for atomic crystals, which then allows discriminating amorphous packing, polycrystalline or single crystals. Depending on the light spot used, the desired spatial resolution might be obtained. The spectrum in Figure 6.9a is an example for a low crystalline quality due to numerous differently oriented microcrystals [180]. However, the spectrum of such a polycrystal can be calculated by superposing Bragg reflections from different orientations of microcrystals with a given crystal structure [89]. In comparison Figure 6.9b shows a spectrum of a colloidal crystal of high crystalline quality [8]. Growing colloidal crystals by sedimentation is a time-consuming procedure, which usually results in a non-adequate crystalline structure as well. As already mentioned above (Figure 6.8) external fields might enhance the crystalline structure, which will be discussed now.

6.3.2.2. External Fields.   The precise control of the sedimentation velocity requires additional external fields, which partly compensate gravity and buoyancy. In literature there are numerous examples where electric [111,164,181–183], magnetic [101,103,105,128,182,184–187] or gravity [125,163,164,167,179,188,189] fields were applied.

When an electric field is applied along the gravitation field the velocity of sinking spheres is simply given by Equation (6.11), which is now extended by the electrical mobility μ of the spheres in an electric field E [190]:

$$v = \frac{d^2 g \left(\rho_{\mathrm{s}} - \rho_{\mathrm{h}}\right)}{2\eta} + \mu E. \tag{6.12}$$

The electrical mobility of the silica spheres can be tuned within certain limit by the pH of the solvent [130]. If the monodisperse SiO_2 nanospheres are of diameters less than 300 nm or more than 550 nm, a controlled crystal growth is usually not successful [130]. If the spheres are too small, the sedimentation rate is very slow or even may not occur. If they are large enough, their sedimentation velocity is such that it is quite hard to achieve an ordered array, and it becomes completely impossible if the diameter is further increased. It is known that when the gravitational energy is much larger than the thermal energy ($k_{\mathrm{B}} T$), the sedimentation occurs far from equilibrium and noncrystalline sediment is obtained. The application of electric fields on charged particles (electrophoresis) [130,164,191], or on uncharged particles (dielectrophoresis) [192–194], allows a precise control of the sedimentation velocity and was investigated in detail by several authors.

Colloidal crystal formation can also be directed by magnetic forces [101,195]. Mainly magnetic fields are used for arraying composite magnetic core-shell particles, where the shell consists of magnetite (Fe_3O_4) nanoparticles with a typical diameter of 10 nm. Because of their single domain nature, these particles bear a permanent magnetic dipole. The energy of one such dipole in the field of a strong permanent magnet is large enough to align these dipoles almost completely with the field. Moreover, the magnetic attraction energy between two magnetic particles is expected to have a noticeable influence on the phase behaviour of a colloidal suspension.

If the silica spheres are too small, resulting in an unacceptable small sedimentation velocity, additional forces are needed in order to reduce the time required for crystal growth. Besides applying an electrical field, moderate acceleration in a centrifugal field is an alternative [118,126,127]. The crystal quality obtained can be optimized by additionally applying ultrasonic. The acoustic field, either phase, frequency or amplitude modulated, acts as a "macroscopic" temperature and supports diffusion during acceleration forced sedimentation. It should be noted that ultrasonic density modulation of the suspension should act only outside the solid phase; otherwise the solid phase would be "heated" again.

6.3.2.3. Self-Assembly by Capillary Forces. So far the forces and methods described have been mainly used to grow bulk crystals. For the use of photonic applications thin opal layers might be sufficient. Thin opal films can be deposited on a flat substrate by a controlled evaporation of the solvent containing a diluted solution of silica spheres. Experimentally this can be achieved either due to lowering the solvent level by cautiously evaporating the solvent or by slowly lifting the substrate such that at the solvent surface a moving meniscus is induced between the spheres (dip coating). However, the quality of the opaline layers received depends strongly on the relative densities of the solvent and the silica spheres, because when using this method sedimentation should be restricted. One is then left with forces acting between the particles only. At the surface of the colloidal suspension capillary forces, strictly speaking the minimization of the surface energy, result in an attractive force between the spheres [196,197]. The sum of attractive capillary forces acting between the spheres is large compared to the forces between particles described above. Since the surface of the solvent between the spheres is enlarged due to the bend meniscus, the evaporation is enhanced between the spheres, thus a convective flux into the direction of the interacting spheres attracts further particles. This results in self-organized growing clusters of silica spheres at the solvent surface [196,198–203]. By controlling the solvent evaporation rate the flux of the particles moving to the growing colloidal clusters can be controlled. Thus the crystalline quality can be optimized by critically choosing an appropriate balance between particle diffusion and crystallization. As one would expect the particle flux as well as the evaporation of the solvent also depends on the temperature. Additional external fields such as ultrasonic or electrical fields can be used to optimize the crystal structure. Nevertheless, this method is mainly used to grow thin colloidal layers, at least monolayers of silica spheres. Besides the probability of moving the solvent surface by evaporation or moving the substrate, in principle, there are two further choices concerning the orientation of the substrate, namely vertical [190,204] and horizontal orientations. Apart from better crystallinity the vertical alignment allows us to vary the number of monolayers simply by changing the bead concentration in the suspension [204]. Application of alternating external force during the crystallization of the colloid, particularly, shear alignment, has been suggested to enhance the crystallinity and to produce large-area ($\sim$cm^2) defect-free opaline crystals [205]. There are a few experiments to further improve the crystal quality [205] using the vertical deposition method, such as experiments on the effect of the evaporation temperature [206]. The growth of colloidal crystals from aqueous solution can be proceeded at a higher evaporation temperature than that from an ethanolic solution; however, the crystal growth rate is lower. It is found that colloidal crystals grown at a higher evaporation temperature ($\sim$>55 °C) show an increasing tendency towards an equilibrium fcc phase, and also have fewer dislocations and vacancies [206].

Thick layers or bulk colloidal crystals are also obtained with a multiple dip coating. Recently, the capability and feasibility of this method in forming 3D opal lattices with well-controlled numbers of layers along the [111] direction were demonstrated [190].

6.3.3. Alignment Under Physical Confinement

Monodispersed colloidal spheres often organize themselves into a highly ordered 3D structure when they are subjected to a physical confinement [173,207,208]. Recently with such a method 3D opal arrays of colloidal spheres could be produced with domain sizes as large as several square centimetres [209–211]. Colloidal spheres (regardless of the forces between spheres, see above) were assembled by gas pressure into a highly ordered structure in a specially designed packing cell. A continuous sonication was the key to the success of this relatively fast method.

Generally for the fabrication of more controlled crystalline arrays, self-assembly of colloidal spheres on a patterned substrate represents a promising method [212]. A patterned substrate has two kinds of geometrical parameters: the periodicity of the pattern and the surface modulation depth or height of the pattern. Several research groups reported on the effects of the periodicity of the pattern on the crystal array [213,214]. It was demonstrated that colloidal crystallization is greatly affected by the ratio of the surface modulation depth or height to the colloid diameter [188]. Especially there is a large effect of the colloidal array on an imprinted substrate [215].

The physical confinement approach combines templating and attractive inter-sphere forces to self-assemble monodispersed nanospheres into complex aggregates with well-controlled sizes, shapes and internal structures [212,215–217]. For example, the use of patterned arrays of relieves on solid substrates to grow colloidal crystals [188,213,218–220] having unusual crystalline orientations and/or structures [221] was demonstrated. Other examples are patterned monolayers directing the deposition of colloidal particles onto designated regions on a solid substrate [215,222,223].

6.4. THREE-DIMENSIONAL PERIODIC NANOPOROUS MATERIALS

Highly ordered 3D structures of monodisperse colloidal spheres made by techniques described above are themselves particularly suitable candidates as templates for highly ordered porous structures. The colloidal 3D structure as a template simply serves as a scaffold around which other kinds of materials are synthesized. After removal of the template an inverted structure remains. The porous materials obtained are called "inverted opals" (see above).

Such a template-directed synthesis is a convenient and versatile method for generating nanoporous materials. Templating against opal arrays of colloidal spheres offers a generic route to nanoporous materials that exhibit precisely controlled pore sizes and highly ordered 3D nanoporous structures [114–116,118,209,224] and has been successfully applied to their fabrication from a wide variety of materials, including organic polymers, ceramic materials, inorganic semiconductors and metals [27,114–116,118–124,225,209,226–230]. The fidelity of this procedure is mainly determined by the quality of the colloidal sphere template, kinetic factors such as the filling void spaces in the template and the volume shrinkage of precursors during solidification (usually by drying or heating).

For the realization of such structures several procedures have been developed. After drying the opal array of colloidal spheres, the void spaces among the colloidal spheres are infiltrated with a liquid precursor such as a dispersion of nanoparticles [228], a solution containing an inorganic salt [119], an ultraviolet or thermally curable organic prepolymer [116,117,209], an ordinary organic monomer [27] or a sol–gel precursor to a ceramic material [27,114–116,118,209,224–227]. Subsequent solidification of the precursor and removal of the colloidal spheres results in a 3D nanoporous structure with a highly ordered architecture of uniform air spheres, interconnected to each other by small circular windows. The void spaces among colloidal spheres have also been filled with a variety of materials through electrochemical deposition [121,123] or by chemical vapour deposition (CVD), with which the degree of filling can be accurately controlled. Thus, CVD was used to fill silica crystals with graphite and diamond [224], silicon [231] and germanium [232]. However, a difficulty is the obstruction with material of the outermost channels which provide access to the innermost channels. By low-pressure CVD, which prevented channel obstruction, and highly ordered silica crystals, inverted crystals of silicon were fabricated [233]. Similarly, a liquid phase deposition process can be used if the deposition occurs under mild conditions and results in conformal hydrous films growing the inverted structure [29].

In the next step the colloidal template is removed from its solidified environment by dissolution, evaporation or firing at temperatures of up to 400 °C. This final step often also involves calcination of the material at elevated temperatures in order for the resulting porous framework to densify and crystallize.

Figure 6.10 shows an inverted opal made from a highly ordered silica nanosphere template which was infiltrated with TiO_2 nanocrystals [226]. The TiO_2 nanocrystals were

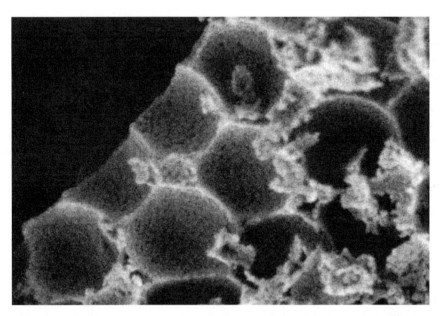

FIGURE 6.10. The inverted opal structure was made from a highly ordered silica nanosphere template infiltrated with TiO_2 nanocrystals, which were grown by a sol–gel process. After infiltration, heating for solidification and removal of the template the inverted structure remains. The size of the TiO_2 nanocrystals is typically up to 10 nm [226].

grown by a sol–gel process [4]. After infiltration, heating for solidification and removal of the template the inverted structure remains, showing the polycrystalline structure of TiO_2 nanocrystals with sizes up to 10 nm (see Figure 6.10).

The fabrication of nanoporous materials based on this approach is not only remarkable for its simplicity, and for its degree of accuracy in transferring 3D nanostructures from the template to the replica, but also a cost-effective and high-throughput procedure. The size of the pores and the periodicity of the porous structures can be precisely controlled and easily tuned by changing the size of the colloidal template spheres. A similar approach is extensible to a large variety of materials. The only requirement seems to be the availability of a precursor that can infiltrate into the void spaces among colloidal spheres without significantly swelling or dissolving the template.

6.5. APPLICATIONS

Colloidal particles have been extensively studied in the context of materials science, condensed matter physics, chemistry, biology, medicine and applied optics [169,234,235]. They have long been used as major components of numerous industrial products such as inks, paints, toners, coatings, foods, drinks, cosmetics or photographic films [149,236,237]. Spherical colloids have been the dominant subject of research for many decades due to their ease of production as monodispersed spheres [238–240] and the wide variety of chemical methods developed to generate them from a range of different materials [6,36,241–243]. Spherical colloids can be readily self-assembled into three-dimensionally ordered lattices such as colloidal crystals or synthetic opals [167,168,244]. The ability to crystallize spherical colloids into spatially periodic structures has allowed us to obtain interesting and often useful functionality not only from the colloidal materials but also from the long-range order exhibited by these crystalline lattices [76,176,177,245]. Until now numerous applications for colloidal particles are known; however, so far only a few applications have been established for colloidal crystals. Materials with spatially ordered features on the nanometre scale have, however, current and future potential applications in optical or magnetical information processing and storage, advanced coatings, catalysis and other emerging nanotechnologies. The long-range ordering of the colloidal particles results in a number of distinctive and potentially useful characteristics such as optical diffraction and photonic bandgaps. Studies on the optical properties of these materials have now evolved into a new and active field of research that is often referred to as photonic crystals or photonic bandgap structures [27,114,116,118,121–124,127,228–231,246,247]. The reader who is interested in the background on the physics of photonic crystals is referred to textbooks and monographs [248–252].

There are also biological applications of colloidal nanocrystals reported [253], for example, the use of nanostructured materials as artificial bones [254], as fluorescent probes to label cells [25,255–263] and chemical libraries [264–266].

Monolayers of magnetic monodisperse core-shell colloids are currently under investigation for their application to ultra-high-density magnetic storage devices [267–270]. Chemical techniques have been used to form silica coatings on nanoparticles of various materials including nickel [271], iron [272–274], iron oxide [275–277] and a nickel/iron composite [274]. Because of the large surface area to volume ratio of nanospheres, coating by magnetic materials is less favourable because metallic nanoparticles are hampered by issues of chemical stability, dispersion and surface functionalization. They are susceptible to attack by oxidative or corrosive environments that may alter their chemistry and diminish their properties, thus they have to be protected by additional coating.

In nanosphere lithography technique the self-assembly of single layer and double layer of nanospheres is used forming a hexagonally close-packed crystal for the deposition of material through this colloidal crystal mask, with subsequent removal of the nanospheres, resulting in an array of evenly spaced, homogeneous nanoparticles [278–281]. The dimension of these truncated periodic nanoparticle arrays can be easily tuned by choice of nanosphere diameter. Nanosphere lithography can be envisioned for applications such as fundamental studies of material properties as a function of particle size, quantum dot arrays [282–284], single-electron transistors and the electrochemistry of nanometre-sized structures (e.g., high-T_c Josephson effect devices [285]).

Dried colloidal crystals are very brittle and may disperse in water. Any practical device thus requires that the crystal either be fixed in place or replicated by another more robust material. Indeed, nature's opals are an example of fixing. A colloidal crystal of silica spheres is made into a solid material after infiltrating the voids between the particles by hydrated silica. Within a few years the fabrication of porous materials using colloidal crystal templates has become a rapidly growing field. Nearly all classes of inorganic and organic materials and metals have been templated into porous ordered structures. The creation of these structures is a fascinating and intellectually challenging problem, but interest in these materials also arises from their wide array of potentially usable applications. One of the most "visible" applications is photonic crystals. Such crystals could increase the efficiency of light emitting diodes and be used in optical processing devices as integrated optical miniature waveguides [286,287], filters and resonators, microcavity lasers [288], mirrors or optical chips [220], thereby acting as analogs to semiconductors in electric circuits [248].

Structured porous materials (inverted opals) can possess desired full photonic bandgaps when created out of a matrix with high refractive index [121,127,228,229, 248,289]. However, such materials have not yet been synthesized, even though recent advances in creating structures via electrochemical growth of semiconductors appear very promising [248].

Apart from photonic crystals and optical applications, the three-dimensional porous materials have potential applications in advanced catalysis, where the hierarchical porosity combines efficient transport and high surface area. Both the bulk and surface chemistry of the materials can be modified to the desired composition [29,118,119]. Interesting catalytic and other applications could be based on the ability of the method to form membranes or composite metallic/dielectric structures. This new field is clearly far from exhausting its store of novel ideas and possibilities [28]. An interesting application could also be photocatalysis (see, e.g., [290]) either when using inverted opals for collecting light or by using directly inverted opals made from semiconductors.

6.6. CONCLUDING REMARKS

Although a variety of preparation methods have been developed, the creation of high quality periodic 3D porous structures, preferably over large areas, uniformly and at low cost, is still a challenging problem. Problems associated with template assisted fabrication of porous structures include preparation of a high quality template, complete filling of the voids in the template and the minimization of shrinkage upon template removal by heating or etching. Since any of these factors can influence the final quality of the porous structure, all these requirements must be fulfilled at the same time.

Devices based on porous titania such as photo-voltaic cells, gas sensors or electrochromic displays have attracted considerable attention in recent years. The efficiency of such devices is enhanced by a precise control of pore size and distribution. Sol–gel fabrication routes offer both low cost and great flexibility in the fabrication of periodic porous 3D structures.

REFERENCES

[1] E. Matijevic, Monodispersed metal (hydrous) oxides—a fascinating field of colloid sciences, Acc. Chem. Res. **14**, 22–29 (1981).

[2] E. Matijevic, Monodispersed colloids: art and science, Langmuir **2**, 12–20 (1986).

[3] C. Feldmann, Preparation of nanoscale pigment particles, Adv. Mater. **13**, 1301–1303 (2001).

[4] C.J. Brinker, *Sol-Gel-Science*, Academic Press, New York, 1990.

[5] H.E. Bergna (Ed.), *Adv. Chem. Series: The Colloid Chemistry of Silica*, Vol. 234, American Chemical Society, Washington, DC, 1994.

[6] W. Stoeber, A. Fink and E. Bohn, Controlled growth of monodisperse silica spheres in the micron size range, J. Colloid Interface Sci. **26**, 62 (1968).

[7] G. De, B. Karmakar and D. Ganguli, Controlled growth of monodisperse silica spheres in the micron size range, J. Mater. Chem. **10**, 2289–2293 (2000).

[8] H. Winkler, Synthese und charakterisierung photonischer bandlückenmaterialien, Diploma thesis, Paderborn, Germany, 2001.

[9] M. Dresselhaus, G. Dresselhaus and P. Avouris, *Carbon Nanotubes*, Springer, Berlin, 2000.

[10] K. Shelimov and M. Moskovits, Composite nanostructures based on template-grown boron nitride nanotubules, Chem. Mater. **12**, 250 (2000).

[11] O. Lourie, C. Jones, B. Bartlett, P. Gibbons, R. Ruoff and W. Buhro, CVD growth of boron nitride nanotubes, Chem. Mater. **12**, 1808 (2000).

[12] R. Ma, Y. Bando, T. Sato and K. Kurashima, Growth, morphology, and structure of boron nitride nanotubes, Chem Mater. **13**, 2965 (2001).

[13] Z. Zhang, B. Wei, J. Ward, R. Vajtai, G. Ramanath and P. Ajayan, Select pathways to carbon nanotube film growth, Adv. Mater. **13**, 1767 (2001).

[14] Y. Sui, D. Acosta, J. Gonzalez-Leon, A. Bermudez, J. Feuchtwanger, B. Cui, J. Flores and M. Saniger, Structure, thermal stability and deformation of multibranched carbon nanotubes synthesized by CVD in the AAO template, J. Phys. Chem. B **105**, 1523 (2001).

[15] A. Peigney, P. Coquay, E. Flahaut, R. Vandenberghe, E. De Grave and C. Laurent, A study of the formation of single- and double-walled carbon nanotubes by a CVD method, J. Phys. Chem. B **105**, 9699 (2001).

[16] A. Morales and C. Lieber, A laser ablation method for the synthesis of crystalline semiconductor nanowires, Science **279**, 208 (1998).

[17] J. Holmes, K. Johnston, R. Doty and B. Korgel, Control of the thickness and orientation of solution-grown silicon nanowires, Science **287**, 1471 (2000).

[18] D. Al-Mawlawi, C.Z. Liu and M. Moskovits, Nanowires formed in anodic oxide nanotemplates, J. Mater. Res. **9**, 1014 (1994).

[19] J.C. Hulteen and C.R. Martin, A general template-based method for the preparation of nanomaterials, J. Mater. Chem. **7**, 1075–1087 (1997).

[20] G. Che, B. Lakshmi, C. Martin, E. Fisher and R. Ruoff, Chemical vapor deposition (CVD)-based synthesis of carbon nanotubes and nanofibers using a template method, Chem. Mater. **10**, 260 (1998).

[21] H. Masuda, T. Yanagishita, K. Yasui, K. Nishio, I. Yagi and A. Fujishima, Synthesis of well-aligned diamond nanocylinders, Adv. Mater. **13**, 247 (2001).

[22] L. Cao, Z. Zhang, L. Sun, C. Gao, M. He, Y. Wang, Y. Li, X. Zhang, G. Li, J. Zhang and W. Wang, Well-aligned boron nanowire arrays, Adv. Mater. **13**, 1701–1704 (2001).

[23] A.P. Alivisatos, Perspectives on the physical chemistry of semiconductor nanocrystals, J. Phys. Chem. **100**, 13226–13239 (1996).

[24] C. Wang, M. Shim and P. Guyot-Sionnest, Electrochromic nanocrystal quantum dots, Science **291**, 2390 (2001).

[25] M.J. Bruchez, M. Moronne, P. Gin, S. Weiss and A.P. Alivisatos, Semiconductor nanocrystals as fluorescent biological labels, Science **281**, 2013–2016 (1998).

[26] Y. Vlasov, X. Bo, J. Sturm and D. Norris, On-chip natural assembly of silicon photonic bandgap crystals, Nature **414**, 2425 (2001).

[27] S.A. Johnson, P.J. Ollivier and T.E. Mallouk, Ordered mesoporous polymers of tunable pore size from colloidal silica templates, Science **283**(12), 963–965 (1999).

[28] O.D. Velev and E.W. Kaler, Structured porous materials via colloidal crystal templating: from inorganic oxides to metals, Adv. Mater. **12**(7), 531–534 (2000).

[29] S. Nishimura, A. Shishido, N. Abrams and T.E. Mallouka, Fabrication technique for filling-factor tunable titanium dioxide colloidal crystal replicas, Appl. Phys. Lett. **81**(24), 4532–4534 (2002).

[30] O.D. Velev and A.M. Lenhoff, Colloidal crystals as templates for porous materials, Curr. Opin. Colloid Interface Sci. **5**, 56–63 (2000).

[31] S. Polarza and B. Smarslya, Nanoporous materials, J. Nanosci. Nanotechnol. **2**, 581–612 (2002).

[32] G.H. Bogush, M.A. Tracy and C.F. Zukoski, Preparation of monodisperse silica particles: control of size and mass fraction, J. Non-Cryst. Solids **104**, 95 (1988).

[33] A. Burneau and B. Humbert, Aggregative growth of silica from an alkoxysilane in a concentrated solution of ammonia, Colloids Surf. A **75**, 111 (1993).

[34] H. Giesche, Synthesis of monodispersed silica powders: I. Particle properties and reaction kinetics, J. Eur. Ceram. Soc. **14**, 189 (1994).

[35] H. Giesche, Synthesis of monodispersed silica powders: II. Controlled growth reaction and continuous production process, J. Eur. Ceram.Soc. **14**, 205 (1994).

[36] F. Candau and R.H. Ottewill, *An Introduction to Polymer Colloids*, Kluwer, Dordrecht, the Netherlands, 1990.

[37] J. Ugelstad, M.S. Elaasser and J. Vanderhoff, The acid decomposition of methylol melamines and methoxymethyl melamines, J. Polym. Sci. Part C: Polym. Lett. **11**, 503–513 (1973).

[38] J.W. Goodwin, J. Hearn, C.C. Ho and R.H. Ottewill, Preparation and characterization of monodisperse polystyrene latexes: III. Preparation without added surface active agents, Colloid Polym. Sci. **252**, 464 (1974).

[39] L. Antl, J.W. Goodwin, R.D. Hill, R.H. Ottewill, S.M. Owens, S. Papworth and J.A. Waters, The preparation of poly(methyl methacrylate) lattices in non-aqueous media, Colloids Surf. **17**, 67 (1986).

[40] C.M. Tseng, Y.Y. Lu, M.S. El Aasser and J.W. Vanderhoff, Uniform polymer particles by dispersion polymerization in alcohol, J. Polym. Sci., Polym. Chem. **24**, 2995 (1986).

[41] G.T.D. Shouldice, G.A. Vandezande and A. Rudin, Practical aspects of the emulsifier-free emulsion polymerization of styrene, Eur. Polym. J. **30**, 179 (1994).

[42] C.E. Reese, C.D. Guerrero, J.M. Weissman, K. Lee and S.A. Asher, Synthesis of highly charged, monodisperse polystyrene colloidal particles for the fabrication of photonic crystals, J. Colloid Interface Sci. **232**, 76 (2000).

[43] C. Feldmann and H.-O. Jungk, Polyol-vermittelte Präparation nanoskaliger Oxidpartikel, Angew. Chemie **113**, 372–374 (2001).

[44] C.R. Silva and C. Airoldi, Acid and base catalysts in the hybrid silica sol-gel process, J. Coll. Interf. Sci. **195**, 381–387 (1997).

[45] A. van Blaaderen and A. Vrij, Synthesis and characterization of monodisperse colloidal organo-silica spheres, J. Colloid Interface Sci. **156**, 1–18 (1993).

[46] A. van Blaaderen and A.P.M. Kentgens, Particle morphology and chemical microstructure of colloidal silica spheres made from alkoxisilanes, J. Non-Cryst. Solids **149**, 161–178 (1992).

[47] T. Matsoukas and E. Gulari, Monomer-addition growth with a slow initiation step: a growth model for silica particles from alkoxides, J. Colloid Interface Sci. **132**, 13–21 (1989).

[48] T. Matsoukas and E. Gulari, Dynamics of growth of silica particles from ammonia-catalyzed hydrolysis of TEOS, J. Colloid Interface Sci. **124**, 252–261 (1988).

[49] G.H. Bogush and C.F. Zukoski, Studies of the kinetics of the precipitation of uniform silica particles through the hydrolysis and condensation of silicon alkoxides, IV, J. Colloid Interface Sci. **142**, 1–18 (1991).

[50] T. Okubo, K. Kobayashi, A. Kuno and A. Tsuchida, Kinetic study of the formation reaction of colloidal silica spheres in microgravity using aircraft, Colloid Polym. Sci. **277**, 483–487 (1999).

[51] J.K. Bailey and M.L. Mecartney, Formation of colloidal silica particles from alkoxides, Colloids Surf. **63**, 131–138 (1992).

[52] S.L. Chen, P. Dong and G.-H. Yang, The size dependence of growth rate of monodisperse silica particles from tetraalkoxysilane, J. Colloid Interface Sci. **189**, 268–272 (1997).

[53] C.J.J. den Ouden and R.W. Thompson, The size dependence of growth rate of monodisperse silica particles from tetraalkoxysilane, J. Colloid Interface Sci. **143**, 77–84 (1991).

[54] K.P. Velikov and A. van Blaaderen, Synthesis and characterization of monodisperse core-shell colloidal spheres of zinc sulfide and silica, Langmuir **17**, 4779–4786 (2001).

[55] C. Blum, Ph.D. thesis, Paderborn, Germany, 2004.

[56] F. Baumann, M. Schmidt, B. Deubzer, M. Geek and J. Dauth, On the preparation of organosilicon m-spheres: a polycondensation in m-emulsion, Macromolecules **27**, 6102–6105 (1994).

[57] D.B. Zhang, H.M. Cheng, J.M. Ma, Y.P. Wang and X.Z. Gai Synthesis of silver-coated silica nanoparticles in nonionic reverse micelles, J. Mater. Sci. Lett. **20**, 439–440 (2001).

[58] J. Esquena, Th. F. Tadros, K. Kostarelos and C. Solans, Preparation of narrow size distribution silica particles using microemulsions, Langmuir **13**, 6400–6406 (1997).

[59] R.I. Nooney, D. Thirunavukkarasu, Y. Chen, R. Josephs and A.E. Ostafin, Synthesis and nanoscale mesoporous silica spheres with controlled particle size, Chem. Mater. **14**(11), 4721–4728 (2002).

[60] Q. Cai, F.Z. Cui, X.H. Chen, Y. Zhang and Z.S. Luo, Nanosphere of ordered silica MCM-41 hydrothermally synthesized with low surfactant concentration, Chem. Lett. **2000**, 1044–1045 (2000).

[61] L. Wang, S. Velu, S. Tomura, F. Ohashi and K. Suzuki, Synthesis and characterization of CuO containing mesoporous silica spheres, J. Mater. Sci. **37**, 801–806 (2002).

[62] A.K. van Helden and A. Vrij, Contrast variation in light scattering: silica spheres dispersed in apolar solvent mixtures, J. Colloid Interface Sci. **76**, 418–433 (1980).

[63] A.K. van Helden, J.W. Jansen and A. Vrij, Preparation and characterization of spherical monodisperse silica dispersions in nonaqueous solvents, J. Colloid Interface Sci. **81**, 354–368 (1981).

[64] K. Bridger, D. Falkenhurst and B. Vincent, Nonaqueous silica dispersions stabilized by terminally-grafted polystyrene chains, J. Colloid Interface Sci. **68**, 190–195 (1979).

[65] H. de Hek and A. Vrij, Preparation of sterically stabilized silica dispersions in nonaqueous media, J. Colloid Interface Sci. **79**, 289–294 (1981).

[66] A. Imhof, M. Megens, J.J. Engelberts, D.T.N. de Lang, R. Sprick and W.L Vos, Spectroscopy of fluorescein (FITC) dyed colloidal silica spheres, J. Phys. Chem. B **103**, 1408–1415 (1999).

[67] A. van Blaaderen and A. Vrij, Synthesis and characterization of colloidal dispersions of fluorescent, monodisperse silica spheres, Langmuir **8**, 2921–2931 (1992).

[68] S. Kang, S. Il Hong, C.R. Choe, M. Park, S. Rim and J. Kim, Preparation and characterization of epoxy composites filled with functionalized nanosilica particles obtained via sol-gel process, Polymer **42**, 879–887 (2001).

[69] O.C. Monteiro, A.C.C. Esteves and T. Trindade, The synthesis of SiO_2@CdS nanocomposites using single-molecule precursors, Chem. Mater. **14**, 2900–2909 (2002).

[70] C.E. Moran, G.D. Hale and N.J. Halas, Synthesis and characterization of lanthanide-doped silica microspheres, Langmuir **17**, 8376–8379 (2001).

[71] M.J.A. de Dood, B. Berhout, C.M. van Kats, A. Polman and A. van Blaaderen, Acid-based synthesis of monodisperse rare-earth-doped colloidal SiO_2 spheres, Chem. Mater. **14**, 2849–2853 (2002).

[72] W. Wang and S.A. Asher, Photochemical incorporation of silver quantum dots in monodisperse silica colloids for photonic crystal applications, J. Am. Chem. Soc. **123**, 12528–12535 (2001).

[73] L.M. Liz-Marzán, M.A. Correa-Duarte, I. Pastoriza-Santos, P. Mulvaney, T. Ung, M. Giersig and N.A. Kotov, Core shell nanoparticles and assemlies thereof, in *Handbook of Surfaces and Interfaces of Materials*, Ed. H.S. Nalwa, Academic Press, San Diego, USA, 2001, Chapter 5.

[74] F. Caruso, Nanoengineering of particle surfaces, Adv. Mater. **13**(1), 11–22 (2001).

[75] N.A.M. Verhaegh and A. van Blaaderen, Dispersions of rhodamine labeled silica spheres: synthesis, characterization, and fluorescence confocal scanning laser microscopy, Langmuir **10**, 1427–1438 (1994).

[76] S.Y. Chang, L. Liu and S.A. Asher, Creation of templated complex topological morphologies in colloidal silica, J. Am. Chem. Soc. **116**, 6745–6747 (1994).

[77] L.M. Liz-Marzan, M. Giersig and P. Mulvaney, Synthesis of nanosized gold-silica core-shell particles, Langmuir **12**, 4329 (1996).

[78] C. Graf, W. Schärtl, K. Fischer, N. Hugenberg and M. Schmidt, Dye-labeled poly(organosiloxane) microgels with core-shell architecture, Langmuir **15**, 6170–6180 (1999).

[79] K.P. Velikov, A. Moroz and A. van Blaaderen, Photonic crystals of core-shell colloidal particles, Appl. Phys. Lett. **80**, 49–51 (2002).

[80] W. Strek, P.J. Deren, E. Lukowiak, J. Hanuza, H. Drulis, A. Bednarkiewicz and V. Gaishun, Spectroscopic studies of chromium-doped silica sol-gel glasses, J. Non-cryst. Solids **288**, 56–65 (2001).

[81] S. Ramesh, Y. Cohen, D. Aurbach and A. Gedanken, AFM investigation of the surface topography and adhesion of nickel nanoparticles to submicrospherical silica, Chem. Phys. Lett. **287**, 461–467 (1998).

[82] V.B. Prokopenko, V.S. Gurin, A.A. Alexeenko, V.S. Kulikauskas and D.L. Kovalenko, Surface segregation of transition metals in sol-gel silica films, J. Phys. D: Appl. Phys. **33**, 3152–3155 (2000).

[83] C.F. Song, M.K. Lü, P. Yang, D. Xu and D.R. Yuan, Study on the photoluminescence properties of sol-gel Ti^{3+} doped silica glasses, J. Sol-Gel Sci. Technol. **25**, 113–119 (2002).

[84] S.M. Jones and S.E. Friberg, Charge transfer transitions of copper (II) in drying silicate xerogels, Phys. Chem. Glasses **37**(3), 111–115 (1996).

[85] M. Nofz, R. Stösser, B. Unger and W. Herrmann, The function of paramagnetic iron species in amorphous materials formed by sol-gel method and conventional melting techniques, J. Non-Cryst. Solids **149**, 62–76 (1992).

[86] M.A. Villegas, M.A. Garcia, J. Llopis and J.M. Fernandez Navarro, Optical spectroscopy of hybrid sol-gel coatings doped with noble metals, J. Sol-Gel Sci. Technol. **11**, 251–265 (1998).

[87] B. Friedel, Dotierung von Siliziumdioxid-Kugeln für photonische Anwendungen, Diploma thesis, Paderborn, Germany, 2003.

[88] Y. Sun and Y. Xia, Alloying and dealloying processes involved in the preparation of metal nanoshells through a galvanic replacement reaction, Nano Lett. **3**, 1569–1572 (2003).

[89] S. Greulich-Weber and E. Waldmüller, unpublished results.

[90] L.H. Slooff, M.J.A. de Dood, A. van Blaaderen and A. Polman, Erbium-implanted silica colloids with 80% luminescence quantum efficiency, Appl. Phys. Lett. **76**(25), 3682–3684 (2000).

[91] L.H. Slooff, A. van Blaaderen, A. Polman, G.A. Hebbink, S.I. Klink, F.C.J.M. van Veggel, D.N. Reinhoudt and J.W. Hofstraat, Rare-earth doped polymers for planar optical amplifiers, J. Appl. Phys. **91**(7), 3955–3980 (2002).

[92] C. De Julian Fernandez, C. Sangregorio, G. Mattei, G. De, A. Saber, S. Lo Russo, G. Battaglin, M. Catalano, E. Cattaruzza, F. Gonella, D. Gatteschi and P. Mazzoldi, Structure and magnetic properties of alloy-based nanoparticles silica composites prepared by ion-implantation and sol-gel techniques, Mater. Sci. Eng. C **15**, 59–61 (2001).

[93] Y. Takeda, C.G. Lee and N. Kishimoto, Optical properties of nanoparticle composites in insulators by high-flux 60 keV Cu^- implantation, Nucl. Instrum. Methods Phys. Res. B **190**, 797–801 (2002).

[94] D.L. Griscom, E' center in glassy SiO_2: microwave saturation properties and confirmation of the primary ^{29}Si hyperfine structure, Phys. Rev. B **20**, 1823 (1979).

[95] M. Donbrow (Ed.), *Microcapsules and Nanoparticles in Medicine and Pharmacy*, CRC Press, Boca Raton, FL, 1992, Chapters 6 and 16.

[96] R. Langner, New methods of drug delivery, Science **249**, 1527–1533 (1990).

[97] F. Caruso, R.A. Caruso and H. Mohwald, Nanoengineering of inorganic and hybrid hollow spheres by colloidal templating, Science **282**, 1111–1114 (1998).

[98] E.M.B. de Sousa, A.P.G. de Sousa, N.D.S. Mohallem and R.M. Lago, Copper-silica sol-gel catalysts: structural changes of Cu species upon thermal treatment, J. Sol-Gel Sci. Technol. **26**, 873–877 (2003).

[99] I.L. Lyubchanskii, N.N. Dadoenkova, M.I. Lyubchanskii, E.A. Shapovalov and Th. Rasing, Magnetic photonic crystals, J. Phys. D: Appl. Phys. **36**, R277–R287 (2003).

[100] C. Koerdt, G.L.J.A. Rikken and E.P. Petrov, Faraday effect of photonic crystals, Appl. Phys. Lett. **82**(10), 1538–1541 (2003).

[101] E.L. Bizdoacaa, M. Spasovaa, M. Farlea, M. Hilgendorffb and F. Carusoc, Magnetically directed self-assembly of submicron spheres with a Fe_3O_4 nanoparticle shell, J. Magn. Magn. Mater. **240**, 44–46 (2002).

[102] C.B. Murray, Shouheng Sun, W. Gaschler, H. Doyle, T.A. Betley and C.R. Kagan, Colloidal synthesis of nanocrystals and nanocrystal superlattices, IBM J. Res. Dev. **45**(1), 47–56 (2001).

[103] R. Wiesendanger, M. Bode, M. Kleiber, M. Lohndorf, R. Pascal, A. Wadas and D. Weiss, Magnetic nanostructures studied by scanning probe microscopy and spectroscopy, J. Vac. Sci. Technol. B **15**, 1330 (1997).

[104] S. O'Brien and J.B. Pendry, Magnetic activity at infrared frequencies in structured metallic photonic crystals, J. Phys.: Condens. Matter **14**, 6383–6394 (2002).

[105] A.J. Haes, C.L. Haynes and R.P. Van Duyne, Nanosphere lithography: self-assembled photonic and magnetic materials, Mater. Res. Soc. Symp. **636**, D4.8/1–D4.8/6 (2001).

[106] A. Moroz, Photonic crystals of coated metallic spheres, Europhys. Lett. **50**(4), 466–472 (2000).

[107] Y. Jiang, C. Whitehouse, Jensen Li, Wing Yim Tam, C.T. Chan and Ping Sheng, Optical properties of metallo-dielectric microspheres in opal structures, J. Phys.: Condens. Matter **15**, 5871–5879 (2003).

[108] N. Eradata, J.D. Huanga, Z.V. Vardenya, A.A. Zakhidovb, I. Khayrullinb, I. Udodb and R.H. Baughmanb, Optical studies of metal-infiltrated opal photonic crystals, Synth. Met. **116**, 501–504 (2001).

[109] A.P. Philipse, Solid opaline packings of colloidal silica spheres, J. Mater. Sci. Lett. **8**(12), 1371–1373 (1989).

[110] J.V. Sanders, Colour of precious opal, Nature **204**, 1151–1153 (1964).

[111] J. Aizenberg, P.V. Braun and P. Wiltzius, Patterned colloidal deposition controlled by electrostatic and capillary forces, Phys. Rev. Lett. **84**, 2997 (2000).

[112] T.C. Simonton, R. Roy, S. Komarneni and E. Breval, Microstructure and mechanical properties of synthetic opal: a chemically bonded ceramic, J. Mater. Res.**1**, 667 (1986).

[113] V.N. Bogomolov, S.V. Gaponenko, I.N. Germanenko, A.M. Kapitonov, E.P. Petrov, N.V. Gaponenko, A.V. Prokofiev, A.N. Ponyavina, N.I. Silvanovich and S.M. Samoilovich, Photonic band gap phenomenon and optical properties of artificial opals, Phys. Rev. E **55**, 7619–7625 (1997).

[114] O.D. Velev, T.A. Jede, R.F. Lobo and A.M. Lenhoff, Porous silica via colloidal crystallization, Nature **389**, 447 (1997).

[115] O.D. Velev, T.A. Jede, R.F. Lobo and A.M. Lenhoff, Microstructured porous silica obtained via colloidal crystal templates, Chem. Mater. **10**, 3597 (1998).

[116] S.H. Park and Y. Xia, Dimensionally interconnected spherical pores, Adv. Mater. **10**, 1045–1046 (1998).

[117] B. Gates, Y. Yin and Y. Xia, Fabrication and characterization of porous membranes with highly ordered three dimensional periodic structures, Chem. Mater. **11**, 2827–2836 (1999).

[118] B.T. Holland, C.F. Blanford and A. Stein, Synthesis of highly ordered three-dimensional mineral honeycombs with macropores, Science **281**, 538–540 (1998).

[119] H. Yan, C.F. Blanford, B.T. Holland, M. Parent, W.H. Smyrl and A. Stein, A general chemical synthesis of periodic macroporous metals, Adv. Mater. **11**, 1003–1006 (1999).

[120] O.D. Velev, P.M. Tessier, A.M. Lenhoff and E.W. Kaler, A class of porous metallic nanostructures, Nature **401**, 548 (1999).

[121] P.V. Braun and P. Wiltzius, Electrochemically grown photonic crystals, Nature **402**, 603–604 (1999).

[122] P.N. Bartlett, T. Dunford and M.A. Ghanem, Templated electrochemical deposition of nanostructured macroporous PbO_2, J. Mater. Chem., **12**, 3130–3135 (2002).

[123] P. Jiang, J. Cizeron, J.F. Bertone and V.L. Colvin, Preparation of macroporous metal films from colloidal crystal, J. Am. Chem. Soc. **121**, 7957–7958 (1999).

[124] P. Yang, T. Deng, D. Zhao, P. Feng, D. Pine, B.F. Chmelka, G.M. Whitesides and G.D. Stucky, Hierarchically ordered oxides, Science **282**, 2244–2246 (1998).

[125] H. Miguez, F. Meseguer, C. Lopez, A. Blanco, J. Moya, J. Requena, A. Mifsud and V. Fornes, Control of the photonic crystal properties of fcc-packed submicrometer SiO_2 spheres by sintering, Adv. Mater. **10**, 480–483 (1998).

[126] S. Tsunekawa, Y.A. Barnakov, V.V. Poborchii, S.M. Samoilovich, A. Kasuya and Y. Nishina, Characterization of precious opals: AFM and SEM observations, photonic band gap, and incorporation of CdS nano-particles, Microporous Mater. **8**, 275 (1997).

[127] J. Wijnhoven and W.L. Vos, Preparation of photonic crystals made of air spheres in titania, Science **281**, 802–804 (1998).

[128] M. Trau, D.A. Saville and I.A. Aksay, Field-induced layering of colloidal crystals, Science **272**, 706 (1996).

[129] M. Trau, D.A. Saville and I.A. Aksay, Assembly of colloidal crystals at electrode interfaces, Langmuir **13**, 6375–6381 (1997).

[130] M. Holgado, F. Garcia Santamaria, A. Blanco, M. Ibisate, A. Cintas, H. Miguez, C.I. Serna, C. Molpeceres, J. Requena, A. Mifsud, F. Meseguer and C. Lopez, Electrophoretic deposition to control artificial opal growth, Langmuir **15**, 4701–4704 (1999).

[131] O. Stern, Zur Theore der elektrischen Doppelschicht, Z. Elektrochem. **30**, 508 (1924).

[132] J.N. Israelachvili, *Intermolecular and Surface Forces*, Academic Press, London and New York, 1992.

[133] M. Elimelech, J. Gregory, X. Jia and R.A. Williams, *Particle Deposition and Aggregation: Measurement, Modeling, and Simulation*, Butterworth, Oxford, 1995.

[134] J. Gregory, Interaction of unequal double layers at constant charge, J. Colloid Interface Sci. **51**, 44–51 (1975).

[135] G.M. Bell, S. Levine and L.N. McCartney, Approximate methods of determining the double-layer free energy of interaction between two charged colloid spheres, J. Colloid Int. Sci. **33**, 335 (1970).

[136] A.K. Arora and B.V.R. Tata (Eds.), *Ordering and Phase Transitions in Charged Colloids*, VCH Publishers, New York, 1996.

[137] P. Debye and E. Hückel, Zur Theorie der Elektrolyte, Physik. Zeitschr. **24**, 305–325 (1923).

[138] G.A. Parks, The isoelectric points of solid oxides, solid hydroxides and aqueous hydroxo complex systems, Chem. Rev. **65**, 177–198 (1977).

[139] R.J. Hunter, *Foundations of Colloid Science*, Vol. 1, Oxford University Press, Oxford, 1986.

[140] H.C. Hamaker, London-van der Waals attraction between spherical particles, Physica **4**, 1058–1072 (1937).

[141] W.C.K. Poon and P.B. Warren, Phase behaviour of hard-sphere mixtures, Europhys. Lett. **28**(7), 513–518 (1994).

[142] H. Sonntag and K. Strenge, *Coagulation Kinetics and Structure Formation*, VEB Deutscher Verlag der Wissenschaften, Kluwer Academic/Plenum, New York, 1988.

[143] J. Mahanty and B.W. Ninham, *Dispersion Forces*, Academic Press, New York, 1976.

[144] B.W. Ninham, K. Kurihara and O.I. Vinogradova, Hydrophobicity, specific ion adsorption and reactivity colloids, Surf. A: Physicochem. Eng. Aspects **123–124**, 7–12 (1997).

[145] B.W. Ninham and V. Yaminsky, Ion binding and ion specificity: the Hofmeister effect and Onsanger and Lifshitz theories, Langmuir **13**, 2097–2108 (1997).

[146] B.W. Ninham, On progress in forces since the DLVO theory, Adv. Colloid Interface Sci. **83**, 1–17 (1999).

[147] B.V. Derjaguin and L.D. Landau, Theory of the stability of strongly charged lyophobic sols and of the adhesion of strongly charged particles in solutions of electrolytes, Acta Physicochim. U.S.S.R. **14**, 633 (1941).

[148] E.J. Verwey and J.Th.G. Overbeek, *Theory of the Stability of Lyophobic Colloids*, Elsevier, Amsterdam, 1948.

[149] W.B. Russel, D.A. Saville and W.R. Schowalter, *Colloidal Dispersions*, Cambridge University Press, Cambridge, 1989.

[150] B.A. Pailthorpe and W.B. Russel, The retarded van der Waals interaction between spheres, J. Colloid Int. Sci. **89**(2), 563–566 (1982).

[151] T. Sugimoto, T. Takahashi, H. Itoh, S. Sato and A. Muramatsu, Direct measurement of interparticle forces by the optical trapping technique, Langmuir **13**(21), 5528–5530 (1997).

[152] S. Asakura and F. Oosawa, Interaction between particles suspended in solutions of macromolecules, J. Polym. Sci. **33**, 183 (1958).

[153] A. Vrij, Polymers at interfaces and the interactions in colloidal dispersions, Pure Appl. Chem. **48**, 471–483 (1976).

[154] S. Asakura and F. Oosawa, On interaction between two bodies immersed in a solution of macromolecules, J. Chem. Phys. **22**, 1255–1256 (1954).

[155] Y. Mao, M.E. Cates and H.N.W. Lekkerkerker, Depletion force in colloidal systems, Physica A **222**, 10–24 (1995).

[156] A. Einstein, On the movement of small particles suspended in stationary liquids required by the molecular-kinetic theory of heat (Über die von der molekularkinetischen Theorie der Wärme geforderte Bewegung von in ruhenden Flüssigkeiten suspendierten Teilchen), Annalen der Physik **17**, 123 (1905).

[157] H. de Hek and A. Vrij, Phase separation in non-aqueous dispersions containing polymer molecules and colloidal spheres, J. Colloid Interface Sci. **70**(3), 592–598 (1979).

[158] C. Pathmamanobaran, H. de Hek and A. Vrij, Phase seperation in mixtures of organophilic spherical silica particles and polymer molecules in good solvents, Colloid Polym. Sci. **259**, 769 (1981).

[159] S.M. Ilett, A. Orrock, W.C.K. Poon and P.N. Pusey, Phase behavior of a model colloid-polymer mixture, Phys. Rev. E **51**(2), 1344 (1995).

[160] O.D. Velev, A.M. Lenhoff and E.W. Kaler, A class of microstructured particles through colloidal crystallization, Nature (London) **287**, 2240 (2000).

[161] R.C. Salvarezza, L. Vázquez, H. Míguez, R. Mayoral, C. López and F. Meseguer, Edward-Wilkinson behavior of crystal surfaces grown by sedimentation of SiO$_2$ nanospheres, Phys. Rev. Lett. **77**(22), 4572–4575 (1996).

[162] M.O. Robbins, K. Kremer and G.S. Grest, Phase diagram and dynamics of Yukawa systems, J. Chem. Phys. **88**(5), 3286–3312 (1988).

[163] P.N. Pusey and W. van Megen, Phase behavior of concentrated suspensions of nearly hard colloidal spheres, Nature **320**, 340–342 (1986).

[164] M. Sullivan, K. Zhao, Ch. Harrison, R.H. Austin, M. Megens, A. Hollingsworth, W.B. Russel, Zhengdong Cheng, Th. Mason and P.M. Chaikin, Control of colloids with gravity, temperature, gradients, and electric fields, J. Phys.: Condens. Matter **15**, S11–S18 (2003).

[165] R. Tuinier, J. Rieger and C.G. de Kruif, Depletion-induced phase separation in colloid-polymer mixtures, Adv. Colloid Interface Sci. **103**, 1–31 (2003).

[166] N.M. Dixit and C.F. Zukoski, Kinetics of crystallization in hard-sphere colloidal suspensions, Phys. Rev. E **64**, 041604–041610 (2001).

[167] P. Pieranski, Colloidal crystals, Contemp. Phys. **24**(1), 25–73 (1983).

[168] I. Snook and W. van Megan, Calculation of the wave-vector dependent diffusion constant in concentrated electrostatically stabilized dispersions, J. Colloid Interface Sci. **100**(1), 194–202 (1984).

[169] A.P. Gast and W.B. Russel, Simple ordering in complex fluids, Phys. Today December, 24 (1998).

[170] N. Ise, How and why do like-charged particles in solution repel one another? Ordered regions in dilute solutions of macroions, Angew. Chem. Int. Ed. Engl. **98**(4), 323–334 (1986).

[171] S. Dosho, N. Ise, K. Ito, S. Iwai, H. Kitano, H. Matsuoka, H. Nakamura, H. Okumura, T. Ono, I.S. Sogami, Y. Ueno, H. Yoshida and T. Yoshiyama, Recent study of polymer latex dispersions, Langmuir **9**, 394–411 (1993).

[172] N.A. Clark, A.J. Hurd and B.J. Ackerson, Single colloid crystals, Nature **281**, 57–60 (1979).

[173] P. Pieranski, L. Strzelecki and B. Pansu, Thin colloidal crystals, Phys. Rev. Lett. **50**, 900 (1983).

[174] T. Okubo, Giant colloidal single crystals of polystyrene and silica spheres in deionized suspension, Langmuir **10**, 1695–1702 (1994).

[175] E.A. Kamenetzky, L.G. Magliocco and H.P. Panzer, Structure of solidified colloidal array laser filters studied by cryogenic transmission electron microscopy, Science **263**, 207–210 (1994).

[176] H.B. Sunkara, J.M. Jethmalani and W.T. Ford, Composite of colloidal crystals of silica in poly(methyl methacrylate), Chem. Mater. **6**, 362 (1994).

[177] M. Weissman, H.B. Sunkara, A.S. Tse and S.A. Asher, Thermally switchable periodicities from novel mesocopically ordered materials, Science **274**, 959–960 (1996).

[178] O. Vickreva, O. Kalinina and E. Kumacheva, Colloid crystal growth under oscillatory shear, Adv. Mater. **12**, 110–112 (2000).

[179] J. Zhu, M. Li, R. Rogers, W. Meyer and R.H. Ottewill, STS-73 space shuttle crew, W.Z.B. Russel and P.M. Chaikin, Crystallization of hard sphere colloids in μ-gravity, Nature **387**, 883 (1997).

[180] S. Beyer, Neue Wege zur Kristallisation von Nanopartikeln zu photonischen Kristallen, Diploma thesis, Paderborn, Germany, 2003.

[181] P. Richetti, J. Prost and P. Barois, Two-dimensional aggregation and crystallization of a colloidal suspension of latex spheres, J. Phys. Lett. **45**, L1137–L1143 (1984).

[182] S.-R. Yeh, M. Seul and B.I. Shraiman, Light-modulated electrokinetic assembly of planar colloidal arrays, Nature **386**, 57–59 (1997).

[183] Y. Solomentsev, M. Bohmer and J.L. Anderson, Electrophoretic cepositiion: a hydrodynamic model, Langmuir **13**, 6058 (1997).

[184] R.C. Hayward, D.A. Saville and I.A. Aksay, Electrophoretic assembly of colloidal crystals with optically tunable micropatterns, Nature **404**, 56–59 (2000).

[185] T. Gong and D.W.M. Marr, Electrically switchable colloidal ordering in confined geometries, Langmuir **17**, 2301–2304 (2001).

[186] C. Mio and D.W.M. Marr, Optical trapping for the manipulation of colloidal particles, Adv. Mater. **12**, 917–920 (2000).

[187] C. Mio and D.W.M. Marr, Tailored surfaces using optically manipulated colloidal particles, Langmuir **15**, 8565–8568 (1999).

[188] A. van Blaaderen, R. Ruel and P. Wiltzius, Template-directed colloidal crystallization, Nature (London) **385**, 321 (1997).

[189] K.E. Davis, W.B. Russel and W.J. Glantschnig, Settling suspensions of colloidal silica: observations and x-ray measurements, J. Chem. Soc. Faraday Trans. **87**, 411 (1991).

[190] P. Jiang, J.F. Bertone, K.S. Hwang and V.L. Colvin, Single-crystal colloidal multilayers of controlled thickness, Chem. Mater. **11**, 2132–2140 (1999).

[191] Y.C. Chan, M. Carles, Nikolaus J Sucher, M. Wong and Y. Zohar, Design and fabrication of an integrated microsystem for microcapillary electrophoresis, J. Micromech. Microeng. **13**, 914–921 (2003).

[192] H.A. Pohl, *Dielectrophoresis*, Cambridge University Press, Cambridge, 1978.

[193] T.B. Jones, *Electromechanics of Particles*, Cambridge University Press, Cambridge, 1995.

[194] S.O. Lumsdon, E.W. Kaler, J.P. Williams and O.D. Veleva, Dielectrophoretic assembly of oriented and switchable two-dimensional photonic crystals, Appl. Phys. Lett. **82**(6), 949–951 (2003).

[195] M. Golosovsky, Y. Saado and D. Davidov, Self-assembly of floating magnetic particles into ordered structures—a promising route for the fabrication of photonic bandgap materials, Appl. Phys. Lett. **75**, 4168–4170 (1999).

[196] N.D. Denkov, O.D. Velev, P.A. Kralchevsky, I.B. Ivanov, H. Yoshimura and K. Nagayama, Two-dimensional crystallization, Nature (London) **361**, 26 (1993).

[197] Q.H. Wei, D.M. Cupid and X.L. Xu, Controlled assembly of two-dimensional colloidal crystals, Appl. Phys. Lett. **77**, 1641–1643 (2000).

[198] G. Picard, Fine particle monolayers made by a mobile dynamic thin laminar flow (DTLF) device, Langmuir **14**(13), 3710–3715 (1998).

[199] N.D. Denkov, O.D. Velev, P.A. Kralchevsky, I.B. Ivanov, H. Yoshimura and K. Nagayama, Mechanism of formation of two-dimensional crystals from latex particles on substrates, Langmuir **8**, 3183–3190 (1992).

[200] A.S. Dimitrov and K. Nagayama, Continuous convective assembling of fine particles into morpho-colored two-dimensional arrays, Langmuir **12**, 1303–1311 (1996).

[201] O.D. Velev, N.D. Denkov, V.N. Paunov, P.A. Kralchevsky and K. Nagayama, Direct measurement of lateral capillary forces, Langmuir **9**, 3702 (1993).

[202] S. Rakers, L.F. Chi and H. Fuchs, Influence of the evaporation rate on the packing order of polydisperse latex monofilms, Langmuir **13**, 7121–7124 (1997).

[203] A.S. Dimitrov, T. Miwa and K. Nagayama, A comparison between the optical properties of amorphous and crystalline monolayers of silica particles, Langmuir **15**, 5257–5264 (1999).

[204] P.A. Kralchevsky, N.D. Denkov, V.N. Paunov, O.D. Velev, I.B. Ivanov, H. Yoshimura and K. Nagayama, Formation of two-dimensional colloid crystals in liquid films under the action of capillary forces, J. Phys.: Condens. Matter **6**, A395 (1994).

[205] R. Amos, J.G. Rarity, P.R. Tapster, T.J. Shepherd and S. Kitson, Fabrication of large-area face-centered-cubic hard-sphere colloidal crystals by shear alignment, Phys. Rev. E **61**, 2929 (2000).

[206] Yong-Hong Ye, F. LeBlanc, A. Hache and Vo-Van Truongb, Self-assembling three-dimensional colloidal photonic crystal structure with high crystalline quality, Appl. Phys. Lett. **78**(1), 52–54 (2001).

[207] D.H. Van Winkle and C.A. Murray, Layering transitions in colloidal crystals as observed by diffraction and direct lattice imaging, Phys. Rev. **34**, 562 (1986).

[208] S. Neser, C. Bechinger, P. Leiderer and T. Palberg, Finite size effects on the closest packing of hard spheres, Phys. Rev. Lett. **79**, 2348 (1997).

[209] S.H. Park, D. Qin and Y.X. Xia, Crystallization of meso-scale particles over large areas, Adv. Mater. **10**, 1028–1031 (1998).

[210] S.H. Park and Y. Xia, Crystallization of meso-scale particles over large areas and its application in fabricating tunable optical filters, Langmuir **15**, 266–273 (1999).

[211] B. Gates, D. Qin and Y. Xia, Assembly of nanoparticles into opaline structures over large areas, Adv. Mater. **11**, 466–469 (1999).

[212] Y. Lu, Y. Yin and Y. Xia, A self-assembly approach to the fabrication of patterned 2D arrays of microlenses of organic polymers, Adv. Mater. **13**, 34–37 (2001).

[213] K. Lin, J.C. Crocker, V. Prasad, A. Schofield, D.A. Weitz, T.C. Lubensky and A.G. Yodh, Entropically driven colloidal crystallization on patterned surfaces, Phys. Rev. Lett. **85**, 1770 (2000).

[214] Y.-H. Ye, S. Badilescu, Vo-Van Truong, P. Rochon and A. Natansohn, Self-assembly of colloidal spheres on patterned substrates, Appl. Phys. Lett. **79**, 872–874 (2001).

[215] Y. Yin, Y. Lu and Y. Xia, A self-assembly approach to the formation of asymmetric dimers from monodispersed spherical colloids, J. Am. Chem. Soc. **123**, 771–772 (2001).

[216] Y. Yin and Y. Xia, Self-assembly of monodispersed colloidal spheres into complex aggregates with well-defined sizes, shapes, and structures, Adv. Mater. **13**, 267–271 (2001).

[217] Y. Yin, Y. Lu and Y. Xia, Self-assembly of monodispersed spherical colloids into 1D chains with well-defined lengths and structures, J. Mater. Chem. **11**, 987–989 (2001).

[218] E. Kim, Y. Xia and G.M. Whitesides, Two- and three-dimensional crystallization of polymeric microspheres by micromolding in capillaries, Adv. Mater. **8**, 245–247 (1996).

[219] S.M. Yang and G.A. Ozin, Opal-chips: vectorial growth of colloidal crystal patterns inside silicon wafers, Chem. Commun. 2507–2508 (2000).

[220] G.A. Ozin and M.Y. Yang, Race for the photonic chip, opal-patterned chips, Adv. Funct. Mater. **11**, 1–10 (2001).

[221] Y. Lu, Y. Yin and Y. Xia, Three-dimensional photonic crystals with non-spherical colloids as building blocks, Adv. Mater. **13**(6), 415 (2001).

[222] J. Tien, A. Terfort and G.M. Whitesides, Microfabrication through electrostatic self-assembly, Langmuir **13**, 5349–5355 (1997).

[223] K.M. Chen, X. Jiang, L.C. Kimerling and P.T. Hammond, Selective self organization of colloids on patterned polyelectrolyte templates, Langmuir **26**, 7825–7834 (2000).

[224] B.T. Holland, C.F. Blanford, T. Do and A. Stein, Synthesis of highly ordered three-dimensional macroporous structures of amorphous or crystalline inorganic oxides, phosphates and hybrid composites, Chem. Mater. **11**, 795–805 (1999).

[225] M. Deutsch, Y.A. Vlasov and D.J. Norris, Conjugated-polymer photonic crystals, Adv. Mater. **12**, 1176 (2000).

[226] A. von Rhein, Synthese, Dotierung und Analyse von TiO_2-Nanopartikeln für die Photonik, Diploma thesis, Paderborn, Germany, 2003.

[227] J.S. Yin and Z.L. Wang, Template-assisted self-assembling and cobalt doping of ordered meroporous titania nanostructures, Adv. Mater. **11**, 469 (1999).

[228] Y.A. Vlasov, N. Yao and D.J. Norris, Synthesis of photonic crystals for optical wavelengths from semiconductor quantum dots, Adv. Mater. **11**, 165 (1999).

[229] V.N. Subramanian, J.D. Manoharan, D.J. Thorne, and D. Pine, Ordered macroporous materials by colloidal assembly: a possible route to photonic bandgap materials, Adv. Mater. **11**, 1261–1265 (1999).

[230] A.A. Zakhidov, R.H. Baughman, Z. Iqbal, C. Cui, I. Khayrullin, S.O. Dantas, J. Marti and V.G. Ralchenko, Carbon structures with three-dimensional periodicity at optical wavelengths, Science **282**, 897–901 (1998).

[231] A. Blanco, E. Chomski, S. Grabtchak, M. Ibisate, S. John, S.W. Leonard, C. López, F. Meseguer, H. Míguez, J.P. Mondía, G.A. Ozin, O. Toader and H.M. van Driel, Large-scale synthesis of a silicon photonic crystal with a complete three-dimensional bandgap near 1.5 micrometers, Nature **405**, 437–440 (2000).

[232] H. Miguez, E. Chomski, F. Garcia-Santamaria, M. Ibisate, S. John, C. Lopez, F. Meseguer, J.P. Mondia, G.A. Ozin, O. Toader and H.M. van Driel, Photonic bandgap engineering in germanium inverse opals by chemical vapor deposition, Adv. Mater. **13**, 1634–1637 (2001).

[233] C.C. Cheng, A. Scherer, V. Arbet-Engels and E. Yablonovitch, Lithographic band gap tuning in photonic band gap crystals, J. Vac. Sci. Technol. B**14**, 4110–4114 (1996).

[234] C.A. Murray and D.G. Grier, Colloidal crystals, Am. Sci. **83**, 238 (1995).

[235] D.G. Grier (Ed.), *From Dynamics to Devices: Directed Self-Assembly of Colloidal Materials*, a special issue in *MRS Bull.* **23**(10), 21 (1998).

[236] D.H. Everett, *Basic Principles of Colloid Science*, Royal Society of Chemistry, London, 1988.

[237] R.J. Hunter, *Introduction to Modern Colloid Science*, Oxford University Press, Oxford, 1993.

[238] Y. Xia, B. Gates, Y. Yin and Y. Lu, Monodispersed colloidal spheres: old materials with new applications, Adv. Mater. **12**, 693–713 (2000).

[239] E. Matijevic, Uniform inorganic colloid dispersions: achievements and challenges, Langmuir **10**, 8 (1994).

[240] E. Matijestic (Ed.), *Fine Particles*, a special issue in *MRS Bulletin* **14**(12), 18 (1989).

[241] R.K. Iler, *The Chemistry of Silica*, Wiley-Interscience, New York, 1979.

[242] I. Piirma (Ed.), *Emulsion Polymerization*, Academic Press, New York, 1982.

[243] G.W. Poehlein, R.H. Ottewill and J.W. Goodwin (Eds.), *Science and Technology of Polymer Colloids*, Vols. 1 and 2, Proc. of ASI, Bristol, England, 1983.

[244] A.D. Dinsmore, J.C. Crocker and A.G. Yodh, Self-assembly of colloidal crystals, Curr. Opin. Colloid Interface **3**, 5–11 (1998).

[245] J.H. Holtz and S.A. Asher, Polymerized colloidal crystal hydrogel films as intelligent chemical sensing materials, Nature **389**, 829–832 (1997).

[246] R. Mayoral, J. Requena, C. López, S.J. Moya, H. Míguez, L. Vázquez, F. Meseguer, M. Holgado, A. Cintas and A. Blanco, 3D long range ordering of submicrometric SiO_2 sintered superstructures, Adv. Mater. **9**, 257–260 (1997).

[247] I.I. Tarhan and G.H. Watson, Photonic band structure of fcc colloidal crystals, Phys. Rev. Lett. **76**, 315 (1996).

[248] J.D. Joannopoulos, R.D. Meade and J.N. Winn, *Photonic Crystals: Molding the Flow of Light*, Princeton University Press, Princeton, 1995.

[249] J.T. Londergan, J.P. Carini and D.P. Murdock, *Binding and Scattering in Two-Dimensional Systems: Application to Quantum Wires, Waveguides and Photonic Crystals*, Springer, Berlin, 1999.

[250] Photonic crystals and light localization, in *Photonic Crystals and Light Localization in the 21st Century 2001, Proceedings of a NATO Advanced Study Institute*, Crete, Greece, Ed C. Soukoulis, NATO Science Series, Kluwer, Dodrecht, 2000.

[251] K. Sakoda, *Optical Properties of Photonic Crystals*, Springer, Berlin, 2001.

[252] S.G. Johnson and J.D. Joannoupolos, *Photonic Crystals: The Road from Theory to Practice*, Kluwer, Boston, 2002.

[253] W.J. Parak, D. Gerion, T. Pellegrino, D. Zanchet, C. Micheel, S.C. Williams, R. Boudreau, M.A. Le Gros, C.A. Larabell and A.P. Alivisatos, Biological applications of colloidal nanocrystals, Nanotechnology **14**, R15–R27 (2003).

[254] T.A. Taton, Boning up on biology, Nature **412**, 491–492 (2001).

[255] W.C.W. Chan and S. Nie, Quantum-dot bioconjugates for ultrasensitive nonisotopic detection, Science **281**, 2016–2018 (1998).

[256] S. Pathak, S.-K. Choi, N. Arnheim and M.E. Thompson, Hydroxylated quantum dots as luminescent probes for in situ hybridization, J. Am. Chem. Soc. **123**(17), 4103–4104 (2001).

[257] E. Klarreich, Biologists join the dots, Nature **413**, 450–452 (2001).

[258] P.S. Weiss, Nanotechnology: molecules join the assembly line, Nature **413**, 585–586 (2001).

[259] S.J. Rosenthal, I. Tomlinson, E.M. Adkins, S. Schroeter, S. Adams, L. Swafford, J. McBride, Y. Wang, L.J. DeFelice and R.D. Blakely, J. Am. Chem. Soc.**124**, 4586–4594 (2002).

[260] J.K. Jaiswal, H. Mattoussi, J.M. Mauro and S.M. Simon, Long-term multiple color imaging of live cells using quantum dot bioconjugates, Nat. Biotechnol. **21**, 47–51 (2003).

[261] B. Dubertret, P. Skourides, D.J. Norris, V. Noireaux, A.H. Brivanlou and A. Libchaber, An in vivo imaging of quantum dots encapsulated in phospholipid micelles, Science **298**, 1759–1762 (2002).

[262] M.E. Akerman, W.C.W. Chan, P. Laakkonen, S.N. Bhatia and E. Ruoslahti, Nanocrystal targeting in vivo, Proc. Natl Acad. Sci. USA **99**, 12617–12621 (2002).

[263] M.X. Wu, H. Liu, J. Liu, K.N. Haley, J.A. Treadway, J.P. Larson, N. Ge, F. Peale and M.P. Bruchez, Immunofluorescent labeling of cancer marker Her2 and other cellular targets with semiconductor quantum dots, Nat. Biotechnol. **21**, 41–46 (2003).

[264] M. Han, X. Gao, J.Z. Su and S. Nie, Quantum-dot-tagged microbeads for multiplexed optical coding of biomolecules, Nat. Biotechnol. **19**, 631–635 (2001).

[265] S.J. Rosenthal, Bar-coding biomolecules with fluorescent nanocrystals, Nat. Biotechnol. **19**, 621–622 (2001).

[266] A.P. Alivisatos, Less is more in medicine—sophisticated forms of nanotechnology will find some of their first real-world applications in biomedical research, disease diagnosis and, possibly, therapy, Sci. Am. **285**, 66–73 (2001).

[267] S. Sun, C.B. Murray, D. Weller, L. Folks and A. Moser, Monodisperse FePt nanoparticles and ferromagnetic nanocrystal superlattices, Science **287**, 1989–1992 (2000).

[268] S. Onodera, H. Kondo and T. Kawana, Materials for magnetic-tape media, MRS Bull. **21**(9), 35 (1996).

[269] K. O'Grady, R.L. White and P.J. Grundy, Whither magnetic recording, J. Magn. Magn. Mater. **177–181**, 886891 (1998).

[270] R.L. White, R.M.H. New and R.F.W. Pease, Patterned media: a viable route to 50 Gbit/in² and up for magnetic recording, IEEE Trans. Magn. **33**, 990–995 (1997).

[271] G. Ennas, A. Mei, A. Musinu, G. Piccaluga, G. Pinna and S. Solinas, Sol-gel preparation and characterization of Ni-SiO² nanocomposites, J. Non-Cryst. Solids **232–234**, 587 (1998).

[272] G.H. Wang and A. Harrison, Preparation of iron particles coated with silica, J. Colloid Interface Sci. **217**, 203–207 (1999).

[273] M. Ohmori and E. Matijevic, Preparation and properties of uniform coated inorganic colloidal particles: 8. Silica on iron, J. Colloid Interface Sci. **160**, 288 (1993).

[274] S. Hui, Y.D. Zhang, T.D. Xiao, M. Wu, S. Ge, W.A. Hines, J.I. Budnick, M.J. Yacaman and H.E. Troiani, in *Nanophase and Nanocomposite Materials IV, Mater. Res. Soc. Symposium Proceedings*, Warrendale, PA, Vol. 703, Ed. S. Komarneni, R.A. Vaia, G.Q. Lu, J.-I. Matsushita and J.C. Parker, Material Research Society, Warrendale, PA, 2002.

[275] A.P. Philipse, M.P.B.V. Bruggen and C. Pathmamanoharan, Magnetic silica dispersions: preparationand stability of surface-modified silica particles with a magnetic core, Langmuir **10**, 92 (1994).

[276] G.-M. Chow and K.E. Gonsalves (Eds.), *Nanotechnology Molecularly Designed Materials*, American Chemical Society, Washington, DC, 1996, p. 42.

[277] Q. Liu, Z. Xu, J.A. Finch and R. Egerton, A novel two-step silica coating process for engineering magnetic nanocomposites, Chem. Mater. **10**(12), 3936–3940 (1998).

[278] H.W. Deckman and J.H. Dunsmuir, Natural lithography, Appl. Phys. Lett. **41**, 377–379 (1982).

[279] C. Hulteen and R.P. Van Duyne, Nanosphere lithography: a materials general fabrication process for periodic particle array surfaces, J. Vac. Sci. Technol. A **13**, 1553–1558 (1995).

[280] F. Burmeister, C. Schäfle, B. Keilhofer, C. Bechinger, J. Boneberg and P. Leiderer, From mesoscopic to nanoscopic surface structures: lithography with colloid monolayers, Chem. Eng. Technol. **21**, 761 (1998).

[281] F. Burmeister, W. Badowsky, T. Braun, S. Wieprich, J. Boneberg and P. Leiderer, Colloid monolayer lithography—a flexible approach for nanostructuring of surfaces, Appl. Surface Sci. **144–145**, 461 (1999).

[282] H. Fang, R. Zeller and P.J. Stiles, Appl. Phys. Lett. **55**, 1433 (1989).

[283] T. Iwabuchi, C. Chuang, G. Khitrova, M.E. Warren, A. Chavez-Pirson, H.M. Gibbs, D. Sarid and M. Gallagher, Fabrication of GaAs nanometer scale structures by dry etching, Proc. SPIE **1284**, 142 (1990).

[284] M. Green, M. Garcia-Parajo and F. Khaleque, Quantum pillar structures on n^+ gallium arsenide fabricated using "natural" lithography, Appl. Phys. Lett. **62**, 264–266 (1993).

[285] W.D. Dozier, K.P. Daly, R. Hu, C.E. Platt and M.S. Wire, Fabrication of high-T_c Josephson effect devices by natural lithography, IEEE Trans. Magn. **27**(2), 3223–3226 (1991).

[286] R.F. Cregan, B.J. Mangan, J.C. Knight, T.A. Birks, P.S. Russell, P.J. Roberts and D.C. Allan, Single-mode photonic band gap guidance of light in air, Science **285**, 1537–1539 (1999).

[287] P.S.J. Russell, T.A. Birks, J.C. Knight, R.F. Cregan, B. Mangan and J.P. De Sandro, Silica/air photonic crystal fibres, Japan. J. Appl. Phys., Part 1 **37**, 45–48 (1998).

[288] A. Guinier and G. Fournet, *Small-Angle Scattering of X-rays*, Wiley, New York, 1955.

[289] G. Subramania, K. Constant, R. Biswas, M.M. Sigalas and K.M. Ho, Optical photonic crystals fabricated from colloidal systems, Appl. Phys. Lett. **74**, 3933 (1999).

[290] L. Bechger, A.F. Koenderink and W.L. Vos, Emission spectra and lifetimes of R6G dye on silica-coated titania powder, Langmuir **18**(6), 2444–2447 (2002).

II

APPLICATIONS

7

Macroporous Silicon Photonic Crystals

A model system for 2D and 3D photonic crystals

Ralf B. Wehrspohn[1], Joerg Schilling[2]

[1]*Department of Physics, University of Paderborn, D-33095 Paderborn, Germany*
wehrspohn@physik.upb.de
[2]*California Institute of Technology, Pasadena, CA 91125, USA*
schill@caltech.edu

7.1. INTRODUCTION

From the beginning of research on photonic crystals, a major area of investigation concerned two-dimensional (2D) photonic crystals [1]. This was mainly caused by experimental reasons as the fabrication of 3D photonic crystals appeared to be more difficult and cumbersome than that of 2D photonic crystals. Additionally, the calculation of band structures for 2D photonic crystals is less time consuming and a lot of interesting phenomena (e.g., light localization—at least in a plane) can already be studied in 2D photonic crystals. However, an ideal 2D photonic crystal consists of a periodic array of infinitely long pores or rods so that the a structure which approximates this theoretical model has to exhibit very high aspect ratios (ratio between pore/rod length to pore/rod diameter). Using conventional dry etching techniques only structures with aspect ratios up to 10–30 are possible. To avoid the scattering of light out of the plane of periodicity and to reduce the corresponding loss the so-called slab structures were developed and thoroughly investigated [2,3]. In such low-aspect structures, one relies on the guiding of light by the total internal reflection in the third dimension and, consequently, deals with a full 3D problem. On the other hand, Lehmann and Grüning [4,5] as well as Lau and Parker [6] proposed macroporous silicon as a model system for 2D photonic crystals. This system consists of a periodic array of air pores in silicon. The pores are etched in hydrofluoric acid applying a photo-electrochemical dissolution process [7,8]. Using lithographic prestructuring the

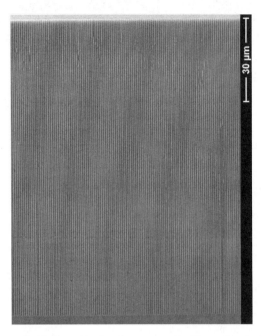

FIGURE 7.1. SEM image of a 2D trigonal lattice of macropores in silicon with a lattice constant of 0.7 μm. As the pore depth amounts to 100 μm the aspect ratio is >100 (courtesy of S. Schweizer).

nucleation spots of the pores can be defined at the surface of the n-type silicon wafer. This also allows us to control the pore pattern and its lattice constant. During the etching process the backside of the wafer must be illuminated to create electronic holes in the silicon which are consumed during the etching process. Due to the electrochemical passivation of the pore walls very high aspect ratios of 100–500 are obtained. As the fundamental band gap appears in general for wavelengths which are approximately twice the lattice constant, the pores are 50–250 times longer than the wavelengths of the corresponding 2D fundamental band gap. Therefore, macroporous silicon represents an excellent system to study ideal 2D photonic crystal properties. In Figure 7.1a structure with a triangular pore lattice with a lattice constant of $a = 700$ nm is shown. The pore depth is 100 μm. In the next paragraphs, optical experiments performed with such structures are presented and compared with calculations assuming a 2D array of infinitely long macropores. The lattice type and the pore depth of the investigated structures are the same as for the sample shown in Figure 7.1 while the interpore distance (lattice constant) and the diameter of the pores varies in order to meet the experimental requirements. Typically, high-quality photonic crystals with lattice constant of $a = 500$–8000 nm can be produced with this process. These structures exhibit photonic bandgaps from the near infrared (IR) to the far infrared.

7.2. 2D PHOTONIC CRYSTALS ON THE BASIS OF MACROPOROUS SILICON

7.2.1. Bulk Photonic Crystals

The dispersion relation for light propagation inside a photonic crystal is calculated using the plane wave method. Due to the 2D periodicity and the uniformity along the third

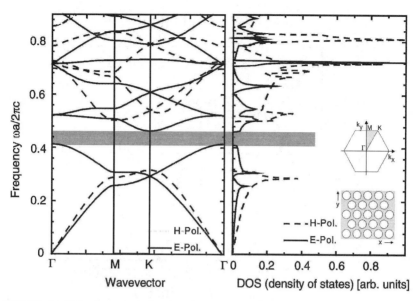

FIGURE 7.2. (a) 2D band structure of a trigonal macroporous silicon photonic crystal ($r/a = 0.45$). (b) Density of photonic states (DOS), inset: 2D hexagonal Brillouin zone and appropriate oriented trigonal pore lattice in real space. The grey bar indicates the 2D complete band gap. In this spectral range neither H-polarized nor E-polarized photonic states exist (DOS = 0) (courtesy of K. Busch).

dimension the light propagating in a 2D photonic crystal splits into E-polarized (E-field parallel to the pore axis) and H-polarized (H-field parallel to the pore-axis) waves. The band structures for these polarizations differ from each other and so do the bandgaps in width and spectral position. This originates in the different field distributions: Typically, the electric field of the H-polarized waves is located in the veins of the structures whereas the electric field of the E-polarized waves concentrates in the connection points of the veins. Figure 7.2a shows an example of a band structure for our system calculated for wave vectors in the first Brillouin zone along the path $\Gamma-M-K-\Gamma$. The assumed porosity or air filling factor is $p = 0.73$ which corresponds to $r/a = 0.45$ (r = pore radius, a = lattice constant) and the refractive index of silicon in the infrared is $n = 3.4$. For a triangular array of pores, a refractive index contrast exceeding 2.7 [9] and for suitable r/a ratios, the bandgaps for E- and H-polarizations overlap and a complete 2D photonic bandgap exists. As the refractive index contrast for air pores in silicon amounts to $n_{Si}/n_{Air} = 3.4$ in the IR, these requirements are fulfilled in our system. The band structure shown in Figure 7.2 thus exhibits such a complete bandgap indicated by a grey bar.

In addition to the band structure, the density of photonic states (DOS) is computed as well and presented in Figure 7.2b [13]. In the spectral region of the complete photonic band gap the DOS is zero, such that the propagation of light in the plane of periodicity with these frequencies is completely forbidden in the photonic crystal. To verify these theoretical calculations, transmission measurements through bars of the macroporous silicon photonic crystals along $\Gamma-M$ and $\Gamma-K$ directions were carried out. For this purpose, bars containing [13] pore rows were cut out using a second lithographic step. The measurements were performed using a Fourier transform infrared spectrometer (FTIR) in the spectral range between 700 cm^{-1} and 7000 cm^{-1} (14.3–1.43 μm). Figure 7.3

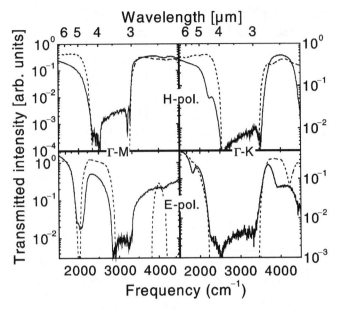

FIGURE 7.3. Transmission measurements (solid) and calculation (dashed) for penetration of a 2D macroporous silicon photonic crystal bar containing 13 pore rows. Transmission for both polarizations (*H*-polarization and *E*-polarization) along the two both high symmetry directions Γ–M and Γ–K are shown [13].

shows the measured spectra for both directions and both polarizations. They are compared to transmission calculations using the method developed by Sakoda [10]. The spectral positions of regions with vanishing transmission correspond well to the calculated spectrum. For the measurements along the Γ–M direction they can be attributed to the band gaps already discussed in Figure 7.2 for *H*-polarized and *E*-polarized light. However, the vanishing transmission in the range of 2200–3500 cm^{-1} for propagation along the Γ–K direction of *E*-polarized light cannot entirely be explained through a stop band. A comparison with the band structure of Figure 7.2 predicts a photonic band which covers the part of this spectral region. However, bands in which the experimentally incident plane wave cannot couple also lead to zero transmission [11,12]. These bands correspond to Bloch modes whose field distributions are antisymmetric with respect to the plane spanned by the pore axis and the direction of incidence. Consequently, although modes do exist in the photonic crystal they need not to be visible in transmission. Care has therefore to be taken when directly comparing reflection or transmission measurements with band structures: Although a band gap always leads to the total reflection/zero transmission, a spectral region exhibiting the total reflection/zero transmission does not necessarily coincide with a band gap. A direct comparison of experiment and theory is therefore rather based on reflection/transmission calculations than on band structure calculations alone. Beside the applied Sakoda method, mainly transfer matrix and finite difference time domain (FDTD) methods have been used for the calculation of reflection and transmission of macroporous silicon photonic crystals (Table 7.1.).

The complete band gap derived from band structure calculations comprises the interval between 2900 and 3300 cm^{-1} (3.44–3.03 µm). It clearly overlaps with all spectral

TABLE 7.1. Methods for the determination of the dispersion relation and the transmission/reflection of photonic crystals used in this work.

Method	Dispersion relation $\omega(k)$	Transmission $I(\omega)$	Reference.
Plane waves	Yes	No	[14]
Sakoda-plane waves	No	Yes	[10,12]
Transfer matrix	No	Yes	[15]
FD-TD	Yes	Yes	[16]

regions with vanishing transmission. The optimum band gap cannot be understood by Bragg scattering only. For scatterers whose spatial dimensions are comparable to the wavelength, additional scattering resonances (known as Mie resonances for spherical particles) appear. They depend on the size and shape of the scatterers. Consequently, apart from symmetry, lattice constant and refractive index, the radius of the pores (r/a-ratio) has an influence on the existence, the position and the width of the photonic band gaps. A graphic representation of the relationship between gap frequencies and filling ratio is known as a gap map, which for our structure, has been calculated before [13]. To verify this gap map experimentally, transmission measurements for 17 different samples spanning a wide range of r/a-ratios were carried out. The band edges were determined from these measurements and are compared with the theoretical predictions in Figure 7.4. The overall correspondence is very good. For lower r/a-ratios only a band gap for the H-polarization exists. A complete band gap only appears for $r/a > 0.4$ as then an E-band gap appears which overlaps with the H-band gap. With increasing r/a-ratios the E-band gap widens while the H-band gap shrinks for very high filling ratios. A maximum complete band gap of $\Delta\omega/\omega = 16\%$ for $r/a = 0.48$ can be deduced. This relatively large complete band gap is a consequence of the strong refractive index contrast between the

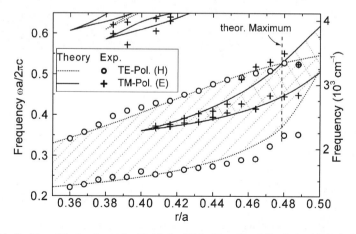

FIGURE 7.4. Position of the band gaps for H-polarized light (dotted) and E-polarized light (solid) for a 2D trigonal macroporous silicon photonic crystal depending on the r/a-ratio (gap map). A complete band gap appears as an overlap of the gaps for both polarizations and attains its maximum size for an r/a-ratio of 0.48 [13].

silicon (pore walls) and air (inside the pores) as well as the synergetic interplay of Mie resonance and Bragg scattering resonance.

7.2.2. Finite Photonic Crystals

Strictly speaking, the band structure calculations can only be performed assuming an infinitely extended photonic crystal. Therefore the band gap (zero DOS) also causing perfect total reflection only appears for infinite bulk photonic crystals. For a very thin bar of the photonic crystal the incident light of a frequency within the bulk band gap is no longer totally reflected. A certain amount can penetrate the thin photonic crystal. To investigate this effect four samples containing 1, 2, 3 and 4 crystal rows with an r/a-ratio of 0.453 were fabricated (Figure 7.5a). Transmission measurements for H-polarized light of different wavelengths along $\Gamma-K$ were performed (see Figure 7.2) [17]. A tunable laser set-up was used which covered the spectral range between $3 < \lambda < 5$ µm corresponding to the range of the H-band gap ($3.1 < \lambda < 5.5$ µm) of the corresponding bulk photonic crystal. The experimental results were compared with transmission calculations applying the already mentioned Sakoda method with 4000 plane waves and revealed a very good agreement (Figure 7.5b). Plotting the transmittance versus the penetrated crystal thickness (Figure 7.5c) an exponential decay is observed. This corresponds to the

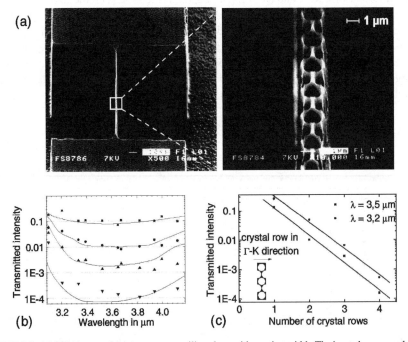

FIGURE 7.5. (a) SEM image of the macroporous silicon bars with varying width. The inset shows an enlarged view of the centre square. (b) Measured and calculated transmission for wavelengths within the H-band gap. Solid lines: Calculations for transmission through 1, 2, 3 and 4 crystal rows. Points: Measurements for 0.89 ± 0.04 (■), 1.8 ± 0.1 (●), 2.9 ± 0.1 (▲) and 4.2 ± 0.2 (▼) crystal rows (determined statistically). (c) Measured transmission as a function of bar thickness for two wavelengths within the band gap [17].

expectation that for frequencies within the band gap the light penetrating into the bulk photonic crystal is exponentially damped. The slope of the line in the logarithmic plot corresponds to a decay constant of 10 dB per crystal row for light with a wavelength near the centre of the band gap. Even for a bar containing only one pore row the band gap is already perceptible. This originates in the strong scattering of the single pores due to the large refractive index contrast between air pores and silicon walls.

7.2.3. Birefringence

In the first years of investigations in photonic crystals mainly the photonic band gap properties were studied. However, over the last years attention was also drawn to the other spectral regions of the dispersion relation that exhibit remarkable properties. For instance, the birefringence of a 2D macroporous silicon photonic crystal has been investigated in the spectral region below the first band gap. From theoretical investigations [18,19], it has been expected that a triangular 2D photonic crystal shows uniaxial properties for $\omega \to 0$. The optical axis coincides in this case with the pore/rod axis. For light propagating in this direction the effective refractive index is independent of the polarization direction (birefringence $= 0$). However, for light propagating in the plane of periodicity the 2D band structure reveals different slopes of the E- and H-polarized bands due to different mode distributions in the silicon matrix. This corresponds to different effective refractive indices for these two different polarizations and leads to birefringent behaviour of light propagation perpendicular to the pore axis. This effect was experimentally investigated in transmission using an FTIR spectrometer. The sample consisted of a macroporous silicon crystal with a lattice constant $a = 1.5$ μm and an r/a-ratio of 0.429. The transmission along Γ–M direction through a bar of 235 μm width containing 181 pore rows was measured [20]. In front of the sample a polarizer was placed and aligned with an angle of $45°$ relative to the pore axis. This defined a certain polarization state of the light incident on the photonic crystal and assured that the radiation consisted of H- and E-polarized components of comparable strengths. After penetration through the sample the beam passes through a second polarizer which was aligned parallel or perpendicular to the first polarizer, respectively. The measured transmission for parallel and crossed polarizers is shown in Figure 7.6. A periodic variation of the transmitted intensity is observed for both polarizer set-ups. The maxima of the parallel polarizer orientation correspond to the minima of the crossed orientation. This can be explained considering the phase difference which builds up between E- and H-polarized lights after penetration through the photonic crystal. This phase difference is given by $\Phi = 2\pi \Delta n_{\text{eff}} df/c$ (Δn_{eff} is the effective refractive index, d is the thickness of penetrated photonic crystal, f is the light frequency). For parallel orientations of the polarizers a maximum occurs for $\Delta \Phi = 2m\pi$ while a minimum appears for $\Delta \Phi = (2m + 1)\pi$. For the crossed polarizers the opposite is true. The light frequency and the order of the maxima and minima are determined from the transmission curve and with this the birefringence Δn_{eff} can be calculated. It is frequency dependent (Figure 7.6). However, over the entire investigated spectral range its value exceeds 0.3 and attains its maximum at the upper limit of the investigated range (at the lower band edge of the first E-gap). The largest birefringence measured amounts to 0.366 at a frequency $f = 0.209c/a$. With this it is by a factor of 43 larger than the birefringence of quartz.

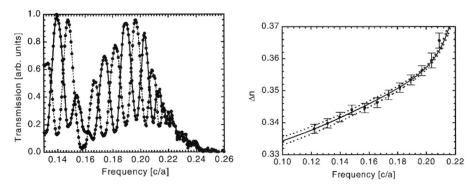

FIGURE 7.6. (Left) Effect of birefringence: measured transmission in the spectral range below the first band gap (long wavelength regime). Spectra were recorded for parallel (solid) and crossed (dashed) orientations of the two polarizers which were placed in front and behind the sample, respectively. The periodic maxima and minima in the transmission spectrum appear due to the phase difference between E- and H-polarized waves accumulating during penetration of the sample [20]. (Right) Spectral dependence of birefringence (Δn). Measurements (data points) and calculations (curves). The dashed curves represent the calculated dependence for the upper and lower bounds of the measured value of r/a (0.429 ± 0.002). The largest measured birefringence ($\Delta n = 0.365$) appears at the upper limit of the investigated spectral range close to the band edge for E-polarization [20].

The uniaxial behaviour of the triangular 2D photonic crystal in the limit $\omega \rightarrow 0$ corresponds to the well-known uniaxial birefringence of hexagonal atomic crystals in the visible. In atomic crystals the scatterers (atoms) have distances in the region of Å and therefore Bragg diffraction occurs for wavelengths in the x-ray region. For these classic atomic crystals, the visible region of the spectrum corresponds to the long wavelength limit $\omega \rightarrow 0$. In our case, where the lattice constant is of the order of 1 μm, Bragg diffraction occurs in the near- and mid-IR (causing the band gaps) while the limit $\omega \rightarrow 0$ comprises the long wavelength regions of the mid- and far-IR. In the described experiment only the birefringence along one propagation direction in the plane of periodicity was investigated. For the case of a uniaxial crystal this is sufficient, as the birefringence is constant for all propagation directions perpendicular to the optical axis. However, for increasing light frequencies which approach the first band gap this is no longer true. In this case, the value of the birefringence depends on the direction of propagation in the Γ–M–K-plane and the optical properties of the crystal can no longer be described by the terms "uniaxial" or "biaxial" known from classic crystal optics [21].

7.3. DEFECTS IN 2D MACROPOROUS SILICON PHOTONIC CRYSTALS

7.3.1. Waveguides

Since the beginning of the study of photonic crystals special attention was paid to intentionally incorporated defects in these crystals. Point or line defects can be introduced into macroporous 2D silicon photonic crystals by omitting the growth of a single pore or a line of pores. This can be achieved by designing a suitable mask for the lithography (the pattern defining process). To demonstrate waveguiding through a linear defect, a 27 μm long line defect was incorporated along the Γ–K direction into a triangular 2D photonic crystal with an r/a-ratio of 0.43 ($r = 0.64$ μm) [22]. However, due to the photo-electrochemical fabrication process, the diameter of the pores, in the adjacent rows to the waveguide is increased.

The transmission through the line defect was measured with a pulsed laser source which was tunable over the whole width of the H-stop band in Γ–K direction (3.1 $< \lambda <$ 5.5 μm). To couple light into the narrow waveguide (with a subwavelength width) with reasonable efficiency, a spatially coherent source of mid-IR light was used. A parametric source was used to produce a beam tunable from 3 to 6 μm, containing 200 fs pulses at a repetition rate of 250 kHz and a typical bandwidth of approximately 200 nm. The H-polarized beam was focused onto the sample by a 19 mm focal length ZnSe lens to a spot size of approximately 25 μm. Because the waveguide width was 1.1 μm, this spot size provided a theoretical coupling efficiency of approximately 4.8%. The transmitted light was passed through a monochromator, chopped, and detected with a pyroelectric detector and a lock-in amplifier. The transmission is defined as the ratio of the transmitted power to the total power incident upon the sample and is about 2%. The transmission deficit compared to 4.8% is attributed to the clipping of the beam by the substrate and diffraction as well as Fresnel losses.

The measured spectrum (Figure 7.7) exhibits pronounced Fabry–Perot resonances over a large spectral range which are caused by multiple reflections at the waveguide facets. Comparing the spectrum with an FDTD transmission calculation reveals very good agreement and the comparable finesse of the measured and calculated resonances indicate small losses inside the sample.

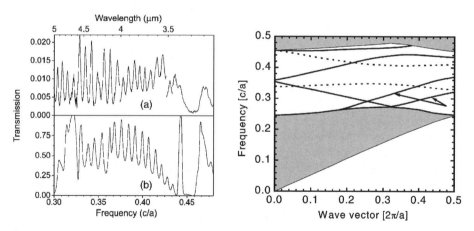

FIGURE 7.7. (Left) (a) Measured and (b) calculated H-polarized transmission spectrum of a 27 μm long waveguide directed along Γ–K covering the spectral range of the H-band gap of the surrounding perfect photonic crystal. The transmission is in %. Only the even waveguide modes contribute to the transmission as the incoming plane wave cannot couple to the odd waveguide modes. The small stop gap at a frequency of 0.45 c/a is caused by the anticrossing of two even waveguide modes [22]. (Right) Computed H-polarized band structure of the waveguide oriented along Γ–K. Solid and dotted curves correspond to even and odd modes, respectively. The two bands which are labelled with arrows appear due to the overetched pores on either side of the waveguide. The shaded areas correspond to the modes available in the adjacent perfect crystal regions [22].

A band structure calculation for *H*-polarization along *Γ*–*K* including waveguide modes is depicted in Figure 7.7. Here, the 2D band structure has been projected onto the new 1D Brillouin zone in *Γ*–*K* direction, since the line waveguide reduces the symmetry. The grey-shaded regions represent all possible modes inside the perfect crystal areas adjacent to the line defect. Defect modes bound to the line defect, therefore, occur only in the band gap, i.e., in the range $0.27 < f < 0.46$. They split into even and odd modes with respect to the mirror plane which is spanned by the waveguide direction and the direction of the pore axis. As the incoming wave can be approximated by a plane wave, the incident radiation can only couple to the even modes of the waveguide. The odd modes do not contribute to the transmission through the waveguide and, therefore, in this experiment transmission is solely connected with the even modes. The small stop band between the even modes around a frequency of 0.45 is reproduced as a region of vanishing transmission in Figure 7.7 due to anticrossing of the waveguide modes [23]. Furthermore, from the band structure it can be concluded that for $0.37 < f < 0.41\ c/a$ only a single even mode exists. Its bandwidth amounts to 10%.

7.3.2. Microcavities

Besides line defects also point defects, consisting only of one missing pore, are of special interest. Such a microresonator-type defect also causes photonic states whose spectral positions lie within the band gap of the surrounding perfect photonic crystal. The light fields belonging to these defect states are therefore confined to the very small volume of the point defect resulting in very high energy densities inside the defect volume. As the point defect can be considered as a microcavity surrounded by perfect reflecting walls, resonance peaks with very high *Q*-values are expected in the transmission spectra. Since the symmetry is broken in both high-symmetry directions, a band structure cannot be used anymore to describe point defect. To study this experimentally, a sample was fabricated including a point defect which was placed between two line defects serving as waveguides for coupling light in and out [24]. Figure 7.8 shows an SEM image of the described sample with an $r/a = 0.433$.

Measuring transmission through this waveguide–microresonator–waveguide structure demands an optical source with a very narrow line width. Therefore, a continuous wave optical parametric oscillator (OPO) has been used which is tunable between 3.6 and 4 μm and delivers a laser beam of 100 kHz line width. For spatially resolved detection an uncoated tapered fluoride glass fibre mounted to an SNOM head was applied and positioned precisely to the exit facet of the outcoupling photonic crystal waveguide (Figure 7.8). In the transmission spectrum two point defect resonances at 3.616 μm and 3.843 μm could be observed (Figure 7.9). Their spectral positions are in excellent agreement with the calculated values of 3.625 μm and 3.834 μm predicted by 2D-FDTD calculations taking into account the slightly widened pores surrounding the point defect. The measured point defect resonances exhibited *Q*-values of 640 and 190, respectively. The differences to the theoretical predicted values of 1700, 750 originate from the finite depth not considered in 2D calculations and the exact pore shape near the cavity. Recent 3D-FDTD calculations show that for high *Q*-values, the finite depth as well as the shape of the pores near the cavity play an important role in the determination of the *Q*-value [25]. Therefore, the 2D limit breaks for high-*Q* cavities under realistic conditions. Intuitively,

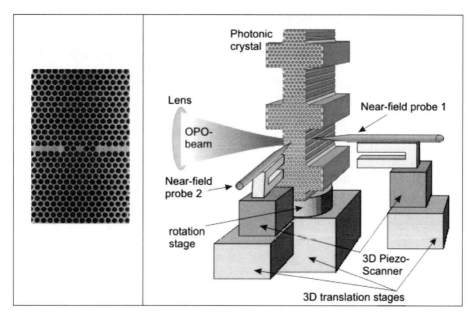

FIGURE 7.8. (Left) Top view of the photonic crystal region containing the waveguide–microresonator– waveguide structure. The r/a-ratio of the pores amounts to 0.433. The waveguides on the left and on the right serve to couple the light into the point defect (microresonator) [24]. (Right) Set-up of the optical measurement (courtesy of V. Sandoghdar).

this can be explained as follows. Any out-of-plane component of the incoming light will result in a spreading of the mode with depth and to a reduction of the Q-value. However, the reported high Q-values of this 2D microresonator might already be sufficient for studying the modification of radiation properties of an emitter placed in such a point defect.

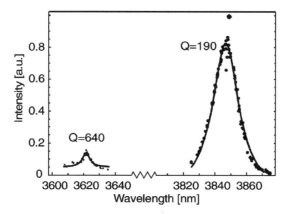

FIGURE 7.9. Measured monopole ($Q = 647$) and decapole resonances ($Q = 191$) of the point defect at wavelengths 3.616 μm and 3.843 μm [24].

7.4. 2D PHOTONIC CRYSTALS IN THE NIR

In the preceding paragraphs, experiments were reported which demonstrate the properties of macroporous silicon for 2D photonic crystals with band gaps in the mid-IR. Their high accuracy makes them a perfect model system to explore the concept of photonic crystals in the IR. Beside their physically interesting properties photonic crystals bear considerable potential for optical telecommunication (for instance, application of line defects for routing of the light beams). For these applications the photonic crystal waveguides have to work in a wavelength range between 1.3 μm and 1.5 μm so that they are compatible to the existing glass fibre network. This fact requires photonic crystals with band gaps in the corresponding spectral range. As it is known from Maxwell's equations, the spectral position of the band gap scales linearly with the lattice constant of the photonic crystals. Therefore, structures with sub-micrometre dimensions are necessary. Although they should not show a novel physical behaviour, their fabrication still is an experimental challenge. A triangular lattice was fabricated with a pitch $a = 0.7$ μm and an r/a-ratio of 0.365. To check the spectral position of the first-order band gap reflection measurements were performed using an IR microscope connected to an FTIR spectrometer. The reflection for H- and E-polarized lights incident in Γ–M direction was measured separately. A gold mirror was used as a reference. Figure 7.10 shows a comparison of the measured reflection spectra with the band structure. The grey-shaded spectral ranges represent the theoretically expected regions of high reflectivity stemming from the band gaps. They correspond very well to the experimental results. Although the reflected light contained contributions from beams with an incidence angle of up to 30° (due to the focussing conditions of the microscope), this off-normal incidence has only a

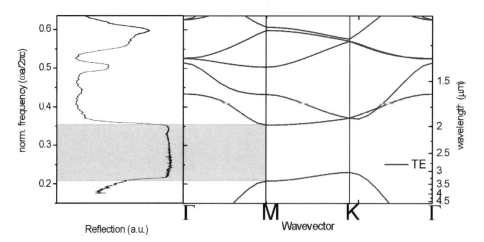

FIGURE 7.10. Reflectivity along Γ–M for a 2D trigonal macroporous silicon photonic crystal with a lattice constant of 0.7 μm for H-polarization (TE). Left: Measured reflectivity of a semi-infinite photonic crystal. Right: Comparison with band structure. Symmetric bands contribute to transmission while for asymmetric bands the incident plane waves cannot couple. Beside the band gaps they also cause total reflection. The dark-shaded range shows the fundamental band gap for H-Pol from 2 to 3.2 μm (courtesy of S. Richter).

negligible effect. The incident light is bent by refraction towards the normal propagating with a much smaller angular deviation inside the photonic crystal. Additionally, the width and position of this first-order band gap is not very sensitive for small angular deviations [26]. Please note that the very steep band edges reflect the very high quality of these structures which were obtained by a recently developed improved etching method. Also reflectivities originating from higher order band gaps, antisymmetric modes or modes with a low group velocity can be observed and are in very good agreement with the theory.

Together with the results of Schilling *et al.* and Rowson *et al.* who showed fundamental band gaps at 1.3 μm [27] and 1.5 μm [28], respectively, this experiment verifies that macroporous silicon structures can be fabricated and used as 2D photonic crystals for the technologically interesting telecommunication wavelengths between 1.3 and 1.5 μm. As was pointed out earlier, the attenuation for light frequencies within the band gap amounts to 10 dB per pore row. As Maxwell's equations scale with the structure size this relative property remains unchanged also for the down-scaled structure. This enables a close packing of waveguides, as the separation of 6–8 pore rows should be sufficient to avoid cross-talk between neighbouring waveguides.

In the last years, surface scattering is getting a more and more severe loss mechanism in III–V semiconductor-based photonic crystals. Silicon has in this sense a unique advantage. Scattering is supposed to originate form surface roughness inside the pores. High dielectric nano-roughness acts as the Mie scatterer for the light. However, high dielectric scatters inside the pores of a silicon photonic crystal can easily be converted to low dielectric scatterers by thermal oxidation. In Figure 7.11, the difference between an as-etched and an oxidized photonic crystal is shown for an extreme case. As-etched macroporous silicon shows in transmission almost no signal in the air band region. After a thin 10 nm oxide has been thermally grown, not only the transmission in the air band increases dramatically, it also enables Fabry–Perot resonances showing the very low losses [29].

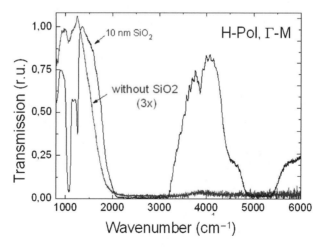

FIGURE 7.11. Effect of a thin oxide coating of the pores of an silicon photonic crystal. In grey, the transmission without the oxide coating is shown, in black the transmission with an oxide coating is shown [29].

7.5. TUNABILITY OF PHOTONIC BAND GAPS

7.5.1. *Liquid Crystals Tuning*

Small deviations of the fabricated experimental structures from designed ones have serious influence on their optical properties. In particular, the design of a microresonator (point defect) with a well-defined resonance frequency in the near-IR allows only fabrication tolerances in the sub-nanometre regime, a demand which currently cannot be fulfilled reproducibly. Additionally, for many applications, e.g., optical switches one would like to shift the band gap during operation. Therefore, tuning the optical properties during operation is a major point of interest. One way, to achieve this behaviour, is to change the refractive index of at least one material inside the photonic crystal. This can be obtained by controlling the orientation of the optical anisotropy of one material incorporated in the photonic crystal [30]. As the proof of the principle of the latter, a liquid crystal (E7 from EM Industries Inc.) was infiltrated into a 2D triangular pore array with a pitch of 1.58 μm and the shift of a band edge depending on the temperature was observed [31]. The liquid crystal E7 is in its nematic phase at room temperature but becomes isotropic at $T > 59$ °C. The refractive index for light polarized along the director axis is $n_e = 1.69$ while it is only $n_0 = 1.49$ for perpendicular polarization exhibiting a strong anisotropy.

Transmission for H-polarized light was measured along the $\Gamma-K$ direction through a 200 μm thick bar of the infiltrated photonic crystal. In the case of room temperature, the first stop band of the H-polarization is observable in the range between 4.4 and 6 μm. Although a large band gap for the H-polarization still exists, the complete band gap, which is characteristic for the unfilled structure, is lost due to the lowered refractive index contrast within the infiltrated crystal. Therefore, the investigations were only carried out for H-polarization. When the structure is heated up, the upper band edge at 4.4 μm is red shifted while the lower band edge exhibits no noticeable shift. At a temperature of 62 °C the red shift saturates and the total shift amounts to $\Delta\lambda = 70$ nm as shown in Figure 7.12. This corresponds to 3% of the band gap width. The shift is caused by the change in the orientation of the liquid crystal molecules inside the pores. In a simplified model one can

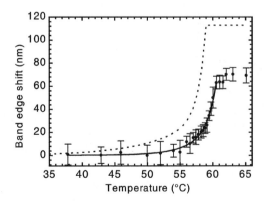

FIGURE 7.12. Temperature dependence of the band edge shift caused by temperature-induced phase transition of the infiltrated liquid crystal. Solid line: Fit to experimental data points. Dashed line: Calculation assuming a simple axial alignment of the liquid crystal in the pores [31].

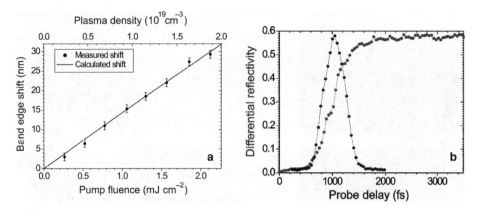

FIGURE 7.13. (a) Band edge shift as a function of the pump fluence, i.e., the plasma density. A maximum shift of 29 nm at 1.9 μm has been observed in good agreement with numerical calculations. (b) Transient behaviour of differential reflectivity at λ = 1900 nm for a pump beam at λ = 800 nm and a fluence of 1.3 mJ/cm². The band edge shift within 400 fs with a dynamic of 25 dB [32].

assume that all liquid crystal molecule directors line up parallel to the pore axis when the liquid crystal is in its nematic phase at room temperature. Then the H-polarized light (E-field in plane) sees the lower refractive index n_0 inside the pores. If the temperature is increased above 59 °C a phase transition occurs and the liquid crystal molecule directors are randomly oriented. The H-polarized light now sees a refractive index inside the pores which is an average over all these orientations. According to this model, a red shift of $\Delta\lambda = 113$ nm is expected which is slightly larger than the measured one. The difference in the observed and calculated shift is currently under investigation. Although the shifting or switching of a band gap via temperature is not very practical for a device, the present study confirms the possible tunability of photonic band gaps using liquid crystals.

7.5.2. Free-Carrier Tuning

Recently, ultrafast tuning of the band edge of a 2D macroporous silicon photonic crystals near 1.9 μm was shown [32]. In contrast to LQ switching, here the refractive index of the silicon matrix was tuned by the optical injection of free carriers. The photonic crystal was illuminated by laser pulse at λ = 800 nm, so well in the absorbtion region of silicon, with a pulse duration of 300 fs. The rise time of the change in the refractive index and thus the shift of the band edge was about 400 fs, slightly slower than the pulse due to the thermalization of the excited carrier (Figure 7.13). The band edge shift observed goes linearly with the pulse intensity as expected from Drude theory. For example, for a pump fluence of 2 mJ/cm², a band shift of 29 nm was observable. This is the fastest switching of 2D photonic crystals to date by free carriers.

7.6. 3D PHOTONIC CRYSTALS ON THE BASIS OF MACROPOROUS SILICON

Thus far, the main work based on macroporous silicon and photonic crystals concerned 2D photonic crystals. However, recently attempts have been undertaken to use

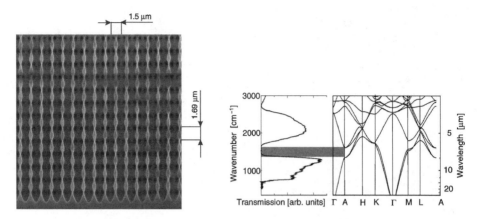

FIGURE 7.14. (Left) SEM image showing a longitudinal section of the modulated pore structure. The variation of the pore diameter with depth can be modelled by a sinusoidal modulation $r = r_0 + \Delta r \sin(2\pi z/l_z)$ with $r_0 = 0.63$ µm, $\Delta r = 0.08$ µm and $l_z = 1.69$ µm. [33]. (Right) Transmission measured in Γ–A-direction (along the pore axis) and comparison with calculated 3D band structure. The grey bar indicates the stop gap for light propagation in this direction causing transmission [33].

macroporous silicon for 3D photonic crystals. One approach to introduce a refractive index variation in the third dimension is the modulation of the pore diameter with pore depth [33]. As described in the first paragraph of this chapter, the pore diameter of the macropores can be controlled during the fabrication process (photo-electrochemical etch process) by the intensity of the backside illumination of the wafer. Strong illumination leads to high etching currents and, therefore, wide pores while the opposite is valid for low illumination.

The illumination intensity was now varied periodically during the etch process applying a zig-zag profile. Figure 7.14 shows an SEM image of a longitudinal section of the sample. The pore diameter modulation can be well approximated by a sinusoidal dependence on the pore depth. The modulation period amounts to 1.69 µm and the porosity varies between 81% and 49% between the planes of wide and narrow pore diameters. The lattice constant a of the 2D pore pattern is again 1.5 µm. The resulting 3D photonic crystal has a hexagonal lattice and the corresponding Brillouin zone has hexagonal shape too. Note that this is the first three-dimensional photonic crystal in the infrared region which perfectly extends over more than 10 lattice periods. To investigate the optical properties of the structure introduced by the pore diameter modulation, reflection measurements were performed along the pore axis which correspond to the Γ–A direction. The spectrum is shown in Figure 7.14 and compared to a 3D band structure calculation using the plane wave method. For comparison with the experiment, the most left part of the band structure shows the relevant dispersion relation along Γ–A. The stop gap in this direction caused by the periodic pore diameter modulation is indicated by a grey bar. It coincides well with the range of zero transmission between 1350 cm^{-1} ($\lambda = 7.41$ µm) and 1680 cm^{-1} ($\lambda = 5.95$ µm) measured along the pores.

Although the structure does not show a complete 3D bandgap it has another distinct property: As it is not based on building blocks of a fixed shape (e.g., spheres or ellipsoids), the periodicity can be different for all directions. The modulation period along the pore axis (z-axis) can be independently controlled from the periodicity in the x–y-plane. Consequently, the dispersion relation along the pores can be adjusted nearly independently from the dispersion relation perpendicular to them. It turned out recently that the same structure but with an initial 2D cubic lattice does have a complete photonic band gap. The resulting structure is an inverted simple cubic lattice with a complete band gap of around 4% for realistic etching parameters [34].

Another approach to fabricate 3D photonic crystals on the basis of macroporous silicon includes a two-step process [35]. In the first step, a conventional 2D array of straight pores is photo-electrochemically etched. Afterwards additional pores are drilled under oblique angles from the top using a focused ion beam (FIB). In this way, a set of three different pore directions is established which cross each other in the depth. The fabricated structure is very similar to the well-known Yablonovite structure for the microwave region. However, a complete 3D band gap could not yet be shown experimentally as the angles between the three different pore sets have not been properly aligned. Another fabrication technique which should give a very similar result uses the photo-electrochemical etching of macropores on a (111) Si surface [36]. In contrast to the pore growth on a (100) Si surface, in the case of a (111) Si surface, the pores grow into ⟨113⟩-directions. As there are three equivalent ⟨113⟩ directions available from the (111) surface, three pores start to grow from each nucleation point at the surface. Band structure calculations for a corresponding structure show that the pores along the three ⟨113⟩ directions grow at suitable angles such that the structure should exhibit a 3D complete photonic band gap of about 7%. Very recently, different crystal structures with complete band gaps larger than 20% based on macroporous silicon have been predicted and are under current investigation [37].

7.7. SUMMARY

In summary, we have reviewed that macroporous silicon is a suitable material to fabricate ideal 2D photonic crystals for the IR. Due to the high refractive index contrast between silicon and air the band gaps are large and for a triangular array of pores a complete band gap for the light propagating in the plane of periodicity appears. Experimental investigations of such a structure for different porosities (r/a-values) confirm the calculated gap map and the maximum width of the complete band gap of 16% for an $r/a = 0.475$. The wide band gap of the H-polarization causes a strong attenuation for light with frequencies within the gap. The corresponding field is exponentially damped and a damping constant of 10 dB per pore row could be experimentally determined. Beside the band gaps also the long wavelength regime below the first band gap has been investigated. Large birefringence was experimentally and theoretically studied and a maximum value of $\Delta n_{eff} = 0.366$ (difference between H- and E-polarizations) was obtained which is by a factor of 43 larger than the birefringence of quartz. Due to the photolithographic prestructuring of the macroporous silicon, defects could intentionally be introduced. The transmission through a straight waveguide has been investigated. After comparison of the experimental features with band structure calculations a single-mode transmission in a spectral range with a bandwidth of 10% could be identified.

Additionally, transmission measurements at a point defect have been performed. Two resonances with Q-values of 647 and 191 were found and comparison with theory reveals that they can be attributed to the monopol and decapol mode of the microresonator. To obtain band gaps in the technologically interesting near-infrared spectral region, macroporous silicon 2D photonic crystals with structure sizes as small as $a = 0.5$ μm have been fabricated. They exhibit band gaps in the optical telecommunication window around $\lambda = 1.3$ μm which was confirmed by reflection measurements. Another issue, closely related to applications, is the tunability of photonic band gaps. A red shift of an upper band edge by 70 nm was demonstrated based on the refractive index change due to the reorientation of liquid crystals infiltrated into the pores. The reorientation was initiated by temperature change and corresponds to the phase transition nematic $\rightarrow$ isotropic of the liquid crystal. Moreover, for the first time the ultrafast switching (400 fs) of the band edge was shown by optically injected free carriers. This speed is compatible with packet switching in telecommunication technology. Finally perfect, extended 3D photonic crystals based on macroporous silicon were presented. Transmission measurements on these 3D photonic crystals with modulated pores showed good agreement with full 3D band structure calculations. Although these photonic crystals do not exhibit a complete 3D band gap the dispersion relation along the pores can almost independently be tuned compared to the dispersion relation perpendicular to it. In particular, one can imagine to utilize the mode structure of these or similar 3D photonic crystals based on macroporous silicon photonic crystals to realize novel atom traps. All these experiments show that macroporous silicon is an ideal material to study the properties of photonic crystals in the infrared regime as well as for possible technological applications operating in this spectral range.

ACKNOWLEDGEMENTS

This chapter is based on the results of a very fruitful collaboration with M. Agio, A. Birner, K. Busch, H.M. van Driel, F. Genereux, U. Gösele, R. Hillebrand, C. Jamois, S. John, P. Kramper, V. Lehmann, S.W. Leonard, J.P. Mondia, F. Müller, S. Richter, V. Sandoghdar, S. Schweizer, C. Soukoulis, and P. Villeneuve. We like to thank all of these. We gratefully acknowledge funding within the Schwerpunkt Programm "Photonische Kristalle" SPP 1113.

REFERENCES

[1] J.D. Joannopoulos, R.D. Meade and J.N. Winn, *Photonic Crystals*, Princeton University Press, New Jersey, 1995.
[2] S.G. Johnson, S. Fan, P.R. Villeneuve, J.D. Joannopoulos and L.A. Kolodziejski, Phys. Rev. B **60**, 5751 (1999).
[3] D. Labilloy, H. Benisty, C. Weisbuch, T.F. Krauss, R.M. De La Rue, V. Bardinal, R. Houdr, U. Oesterle, D. Cassagne and C. Jouanin, Phys. Rev. Lett. **79**, 4147 (1997).
[4] U. Grüning, V. Lehmann and C.M. Engelhardt, Appl. Phys. Lett. **66**, 3254 (1995).
[5] U. Grüning, V. Lehmann, S. Ottow and K. Busch, Appl. Phys. Lett. **68**, 747 (1996).
[6] H.W. Lau, G.J. Parker, R. Greef and M. Hölling, Appl. Phys. Lett. **67**, 1877 (1995).
[7] V. Lehmann and H. Föll, J. Electrochem. Soc. **137**, 653 (1990).
[8] V. Lehmann, J. Electrochem. Soc. **140**, 2836 (1993).

[9] R.D. Meade, K.D. Brommer, A.M. Rappe and J.D. Joannopoulos, Appl. Phys. Lett. **61**, 495 (1992).

[10] K. Sakoda, Phys. Rev. B **52**, 8992 (1995).

[11] W.M. Robertson, G. Arjavalingam, R.D. Meade, K.D. Brommer, A.M. Rappe and J.D. Joannopoulos, Appl. Phys. Lett. **61**, 495 (1992).

[12] K. Sakoda, Phys. Rev. B **52**, 7982 (1995).

[13] A. Birner, A.-P. Li, F. Müller, U. Gösele, P. Kramper, V. Sandoghdar, J. Mlynek, K. Busch and V. Lehmann, Mater. Sci. Semicond. Process. **3**, 487 (2000).

[14] K.M. Ho, C.T. Chan and C.M. Soukoulis, Phys. Rev. Lett. **65**, 3152 (1990).

[15] J.B. Pendry and A. McKinnon, Phys. Rev. Lett. **69**, 2772 (1992).

[16] A. Taflove, *Computational Electrodynamics: The Finite-Difference Time-Domain Method*, Artech House, Boston (1995).

[17] S.W. Leonard, H.M. van Driel, K. Busch, S. John, A. Birner, A.-P. Li, F. Müller, U. Gösele and V. Lehmann, Appl. Phys. Lett. **75**, 3063 (1999).

[18] A. Kirchner, K. Busch and C.M. Soukoulis, Phys. Rev. B **57**, 277 (1998).

[19] P. Halevi, A.A. Krokhin and J. Arriaga, Appl. Phys. Lett. **75**, 2725 (1999).

[20] F. Genereux, S.W. Leonard, H.M. van Driel, A. Birner and U. Gösele, Phys. Rev. B **63**, 161101(R) (2001).

[21] M.C. Netti, A. Harris, J.J. Baumberg, D.M. Whittaker, M.B.D. Charlton, M.E. Zoorob and G.J. Parker, Phys. Rev. Lett. **86**, 1526 (2001).

[22] S.W. Leonard, H.M. van Driel, A. Birner, U. Gösele and P.R. Villeneuve, Opt. Lett. **25**, 1550 (2000).

[23] H. Benisty, S. Olivier, M. Rattier and C. Weisbuch, in Photonic Crystals and Light Localization in the *21st Centruy*, Kluwer Academic Publishers, Dordrecht, 2001, pp. 117–128.

[24] P. Kramper, A. Birner, M. Agio, C. Soukoulis, U. Gösele, J. Mlynek and V. Sandoghdar, Phys. Rev. B **64**, 233102 (2001).

[25] M. Kafesaki and V. Sandoghdar, private communication (2003).

[26] S. Rowson, A. Chelnokov, C. Cuisin and J.-M. Lourtioz , J. Opt. A: Pure Appl. Opt. **1**, 483 (1999).

[27] J. Schilling, A. Birner, F. Müller, R.B. Wehrspohn, R. Hillebrand, U. Gösele, K. Busch, S. John, S.W. Leonard and H.M. van Driel, Opt. Mater. **17**, 7 (2001).

[28] S. Rowson, A. Chelnokov and J.M. Lourtioz, Electron. Lett. **35**, 753 (1999).

[29] A. Birner, *Dissertation*, University Halle, 2000.

[30] K. Busch and S. John, Phys. Rev. Lett. **83**, 967 (1999).

[31] S.W. Leonard, J.P. Mondia, H.M. van Driel, O. Toader, S. John, K. Busch, A. Birner, U. Gösele and V. Lehmann, Phys. Rev. B **61**, R2389 (2000).

[32] S.W. Leonard, H.M. van Driel, J. Schilling and R.B. Wehrspohn, Phys. Rev. B **66**, 161102 (2002).

[33] J. Schilling, F. Müller, S. Matthias, R.B. Wehrspohn and U. Gösele, Appl. Phys. Lett. **78**, 1180 (2001).

[34] S.W. Leonard, Appl. Phys. Lett. **81**, 2917 (2002).

[35] A. Chelnokov, K. Wang, S. Rowson, P. Garoche and J.-M. Lourtioz, Appl. Phys. Lett. **77**, 2943 (2000).

[36] M. Christophersen., J. Carstensen, A. Feuerhake and H. Föll, Mater. Sci. Eng. B **69–70**, 194 (2000).

[37] R. Hillebrand, S. Senz and W. Hergert, J. Appl. Phys. **94**, 2758 (2003).

8

High-Density Nickel
Nanowire Arrays

Kornelius Nielsch[1], Riccardo Hertel[2] and Ralf B. Wehrspohn[3]

[1] *Max-Planck-Institute of Microstructure Physics, Weinberg 2, 06120 Halle, Germany*
knielsch@mpi-halle.de
[2] *Institute of Solid State Research (IFF) Research Center Jülich, D-52425 Jülich, Germany,*
r.hertel@fz-juelich.de
[3] *Department of Physics, University Paderborn, 33095 Paderborn, Germany*
wehrspohn@physik.upb.de

8.1. INTRODUCTION

The application of ferromagnetic-material-filled ordered matrices for perpendicular magnetic storage media is becoming increasingly relevant to extend the areal density of magnetic storage media beyond the predicted superparamagnetic limit (>70 Gbit/in^2) [1,2]. One bit of information corresponds to one single-domain nanosized particle or so-called nanomagnet. Since each bit would be composed of a single large aspect particle, the areal density of pattered media can, in principle, be much more than an order of magnitude higher than that in conventional longitudinal media. For example, an areal density of about 1 Tbit/in^2 can be achieved by a hexagonally arranged array of nanomagnets with a lattice constant of about 25 nm.

The fabrication of nanomagnet arrays based on hexagonally arranged porous alumina as a template material is cheaper than that based on traditional fabrication methods such as nanoscaling using electron beam lithography [3]. Moreover, these arrays of magnetic nanowires can be easily fabricated over areas of several cm^2. Since 1981, several articles on unarranged porous alumina templates filled with ferromagnetic materials have been published [4–8]. These structures have large size distributions of the pore diameter and interpore distance, and the filling degree of the pores is not specified. On the basis of an approach by Masuda [9] (see also Chapter 3), we have shown that ordered porous alumina arrays with a sharply defined pore diameter and interpore distance can be obtained by a two-step electrochemical anodization process of aluminium [10,11]. The degree of self-ordering is polydomain with a typical domain size of a few micrometres.

Monodomain pore arrays can be obtained by electron-beam lithography [12] or imprint technology [13].

8.2. EXPERIMENTAL DETAILS

8.2.1. Preparation of the Porous Alumina Structure

The hexagonally ordered porous alumina membranes were prepared via a two-step anodization process, which is described in detail elsewhere [13,14]. A first long-time anodization caused the formation of channel arrays with a high aspect ratio and regular pore arrangements via self-organization [9–14]. After complete dissolution of the oxide structure (Figure 8.1a), the surface of the aluminium substrate kept the regular hexagonal texture of the self-organized pore tips, which act as a self-assembled mask for a second anodization process. After a second anodization for 1 hour, an ordered nanopore array (Figure 8.1b) was obtained with straight pores from top to bottom and a thickness of typically 1 μm. The parameters are 0.3 M oxalic acid, $U_{ox} = 40$ V and $T = 2\,°$C.

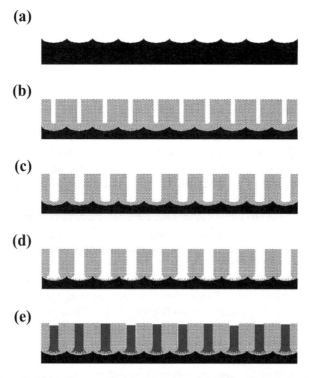

FIGURE 8.1. Schematic diagram demonstrating the fabrication of a highly ordered porous alumina matrix and the preparatory steps necessary for the subsequent filling of the structure. The Al-substrate was pre-structured by a long-time anodizsation and by removing the oxide (a). A second anodizsation step yielded a highly ordered alumina pore structure (b). The barrier layer was thinned and the pores were widened by isotropic chemical etching (c). To thin the barrier layer further, two current-limiting anodizsation steps followed, with dendrite pores forming at the barrier layer (d). Pulsed electrodeposition of nickel in the pores is shown in (e).

Thinning of the barrier layer improves the quality and homogeneity of the deposition process in the pores significantly. The barrier layer can be thinned by chemical pore widening and by current-limited anodization steps: Firstly, the oxalic acid was heated up to 30 °C to decrease the thickness of the barrier layer by chemically widening the pores (Figure 8.1c). After 3 hours, for example after 3h pore widening, the barrier layer was decreased from 45 to 30 nm and the mean pore diameter was increased to approximately 50 nm. Afterwards, the electrolyte was cooled down to 2 °C to interrupt the widening process.

Secondly, the structure was anodized twice for 15 minutes using constant current conditions of 290 and 135 mA/cm^2, respectively. During these anodization steps, the anodization potential decreased, the pores branched out at the formation front and the thickness of the barrier layer was reduced significantly. Finally, the anodizing potential reached a value of 6–7 V, which corresponds to a barrier oxide thickness of less than 10 nm. A detailed description of the pretreatment of the porous alumina structure for the filling process of the pores has been published recently [14].

8.2.2. Filling of the Pores with Magnetic Materials

Nickel and cobalt were electrodeposited from an aqueous electrolyte at the pore tip of our high-aspect-ratio porous material (Figure 8.1e). Both metals were deposited from a highly concentrated Watts-bath electrolyte to achieve a high concentration of metal ions in each pore. The ingredients for nickel electrolyte is given as follows: 300 g/l $NiSO_4 \cdot 6H_2O$, 45 g/l $NiCl_2 \cdot 6H_2O$, 45 g/l H_3BO_3, pH = 4.5. The mixture for the cobalt deposition is written in a similar manner and it has a pH value of 4.3. The electrolyte temperature is 35 °C.

Frequently, an alternating current (ac) signal is used for the deposition [1,2,5,7,15–17] when the porous alumina structure is kept on its aluminium substrate for the filling process. The metal is directly deposited upon the nearly isolating oxide barrier layer at the pore tips. Recently we have demonstrated that a pulsed electrodeposition concept (PED) is more suitable for a direct and homogeneous filling of the porous alumina structures. Here, only a short technical description is given.

The pore filling was based on modulated pulse signals in the ms-range. During the relatively long pulse of negative current (8 ms, $I_{pulse} = -70$ mA/cm^2) the metal is deposited on the pore ground. The measured voltage signal varies between -8 and -12 V. After the deposition pulse, a short pulse of positive polarization (2 ms, $U_{pulse} = +4$V) follows to interrupt the electric field at the deposition interface immediately. The relative long break time (0.3 to 1s) was allowed between the deposition pulses to refresh the ion concentration at the deposition interface, to let disappear the deposition by-products from the pore tips and to ensure a stable pH value in each pore during the deposition. Consequently, the delay time t_{off} improves the homogeneity of the deposition. For the deposition of nickel a $t_{off} = 990$ ms was selected. The deposition was continued up to the beginning of the metal deposition on top of the matrix structure.

8.2.3. Characterization of the Filling Material

For the characterization of the metal-filled template and its magnetic properties, the top of the template structure was fixed to a silicon substrate by conducting glue. Next, the

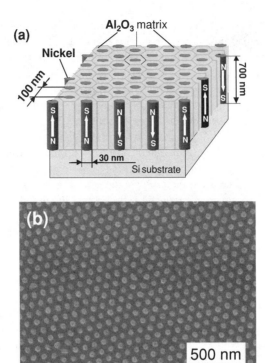

FIGURE 8.2. (a) Sketch of the magnetic structure. Nickel nanowires are arranged in a hexagonal array perpendicular to a silicon substrate and embedded in an aluminium oxide matrix. (b) Top-view SEM micrograph of a nickel-filled alumina matrix, with an interpore distance of 105 nm, fixed on a silicon substrate. The Ni columns have a diameter $D_P = 35$ nm and a length of ∼700 nm.

aluminium substrate was removed by a saturated solution of HgCl and the structure was turned upside down. After removing from the top an ≈200-nm-thick layer of the filled template by a focused ion beam, which was estimated from the thinning rate, the top ends of the nanowires became visible at the surface and a relatively smooth surface was obtained.

As an example, Figure 8.2b shows an SEM image of a nickel sample with a nanowire diameter of 35 nm and 105 nm interpore distance. The ferromagnetic nanowires (white) with a monodisperse diameter are embedded in the porous alumina matrix (black). Because of the self-organization process, the nanowires are arranged in a hexagonal pattern. Figure 8.3 shows an SEM image of sample (B). The ferromagnetic nanowires (white) are embedded in the porous alumina matrix (black). Because of the self-organization process, the nanowires are hexagonally arranged with an interwire distance of 100 nm. Sample (B) has a wire diameter of approximately 35 nm. Nearly 100% pore filling was obtained for all three samples discussed, demonstrating that the metallic filling extends over the whole length of the pore [14]. The crystallinity of these samples was further analyzed by X-ray diffraction (XRD).

From the 2ϑ-scan, the average crystallite size is estimated using the Scherrer equation for round particles, yielding an average grain size $D_{Gr} = 10$–15 nm.

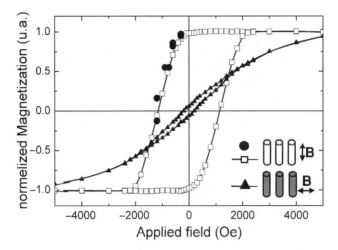

FIGURE 8.3. SQUID-hysteresis loops of the nickel nanowire array with a pitch of 105 nm, a column length of about 700 nm and a wire diameter of 30 nm measured with an applied field parallel (□) and perpendicular (▲) to the column axis. Results from MFM investigations (●) while an external magnetic field H_{ex} was applied to the sample (analyzed statistically treated).

8.3. MAGNETIC PROPERTIES OF NICKEL NANOWIRE ARRAYS

The bulk magnetic properties of the highly ordered array with high-aspect-ratio magnetic columns (length/diameter ~25) were investigated by SQUID-magnetometer measurements. Figure 8.3 shows the bulk magnetization hysteresis loops for the nickel array measured with an applied field parallel and perpendicular to the wire axis.

The hysteresis loops measured for the nickel nanowire (Figure 8.3) array with the magnetic field applied parallel to the wire axis show a coercive field of $H_C^{\parallel} = 1200$ Oe and a squareness of nearly 100%. The measured coercive fields for the hysteresis loops measured perpendicular to the wire are $H_C^{\perp} \approx 150$ Oe and drastically smaller than $H_C^{\parallel}$. This sample has a preferential magnetic orientation along the wire axis. Here, we give only a short description of the analysis of the sample. A detailed analysis of the influence of the nanowire diameter on the bulk-magnetic properties of hexagonally ordered 100 nm period nickel nanowire arrays can be found in [18]. The total magnetic anisotropy of this sample is influenced by the magnetic anisotropy resulting from the shape of the Ni nanowires ($H_S = 2\pi M_S = 3200$ Oe) and the dipole or so-called demagnetization fields ($H_D^{\parallel} = \pi^2 M_S D_P^2 c \approx 550$ Oe; c: pore density) between the nanowires. The magnetocrystalline anisotropy of the nano-crystalline Ni wires gives only a small contribution ($H_K = 2K_1/M_S = 195$ Oe) at room temperature. The Ni nanomagnets are single-domain particles. Their magnetic reversal process occurs by inhomogeneous switching modes, as discussed in the micromagnetic modelling Section 8.5 and [19]. The small size distribution of the pore diameter ($\Delta D_P/D_P < 10\%$) [9,11] has a positive impact on the magnetic properties. Here, we report the highest measured coercive fields H_C of about 1200 Oe for a close-packed nickel nanowire array embedded in a membrane matrix. Previous works on unarranged nickel nanowire arrays show lower coercive fields of about 1000

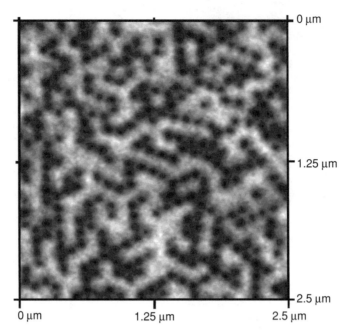

FIGURE 8.4. Magnetic force microscopic image of the highly ordered nickel nanowire array with a pitch of 105 nm embedded in the alumina matrix in the demagnetized state, showing the magnetic polariszation of the pillars alternately "up" (white) and "down" (black).

Oe or less in the preferential magnetic orientation [8,20]. The large size distribution (up to $\Delta D_P/D_P > 50\%$) [7] of the pore diameters and the interwire distance enhance the magnetic interactions in the nanowire arrays and reduce the squareness of the hysteresis loop.

In contrast, the MFM image reflects the magnetic polarization at the top end of each magnetic nanowire. Figure 8.4 demonstrates the domain structure of an array of nickel columns in the demagnetized state. The geometric parameters of the sample are the same as for Figure 8.1b. Dark spots in the magnetic image imply the magnetization pointing up and a bright spots imply the magnetization pointing down. Up magnetization may be interpreted as a binary "1" and down magnetization as a binary "0". It can be deduced from the picture that the Ni pillars are single-domain nanomagnets aligned perpendicular to the surface. The patterned domain structure is due to an antiferromagnetic alignment of pillars influenced by the weak magnetic interaction between these nanomagnets. The labyrinth pattern (Figure 8.4) of the domain structure is characteristic for hexagonally arranged single-domain magnetic particles with a perpendicular magnetic orientation in the demagnetized state. In the case of a quadratic lattice, each of the four nearest neighbours will be aligned anti-parallel and the domain structure exhibits a checkerboard pattern [2]. In the hexagonal lattice, two of the six nearest neighbours will align their magnetization parallel and four will be magnetized anti-parallel, if the stray field has only nearest-neighbour interaction. In Figure 8.4, we observe that in average 2.5 nanomagnets are aligned parallel and that 3.5 are magnetized anti-parallel. We suppose that the stray

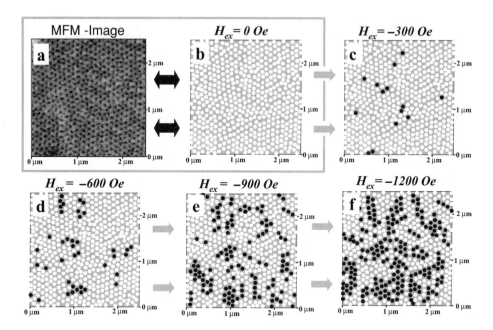

FIGURE 8.5. An external magnetic field was applied perpendicular to the sample surface. (a) The MFM image of the completely magnetizsed nanowire array. The MFM image (a) was enhanced numerically: (a)→(b). Numerically enhanced MFM images recorded with an applied magnetic field of $H_{ex} = -300$ Oe (c), -600 Oe (d), -900 Oe (e) and -1200 Oe (f).

field interaction is extended over several lattice periods D_{int}, due to the high aspect ratio of the magnetic nanowire.

Additionally, MFM investigations with applied magnetic field were carried out on the sample to study the switching behaviour of the individual nanowires in the array. A low moment magnetic tip was used for the MFM scan, in order to prevent switching of the magnetization in the nanowires by the dipole field of the magnetic tip ($H_{tip} \approx$ 50 Oe). Before this investigation, the sample was completely magnetized by an external magnetic field of about 5000 Oe along the wire axes. The first scan was performed without an external field, see Figure 8.5a. In order to get a better impression about the magnetic polarization of each pillar the MFM images were numerically enhanced (Figures 8.5a → and 8.5b). There are no differences in the magnetic polarization between the magnetic pillars (Figure 8.5a) and in the enhanced image every nanowire shows a positive polarization (white dots). Because of the fact that the applied field H_{ex} was larger than the saturation field $H_S^{\parallel}$ and that the hysteresis loop has a magnetic squareness of about 100%, we may deduce that the magnetization in each pillar of the array is oriented in one direction (up or down). Even though $H_{ex} = 0$ Oe and the maximum demagnetization field of $H_D = -550$ Oe is achieved, the structure remains in the saturated state. During the following MFM scans an increased external magnetic field is applied in the direction opposite to the magnetization. The numerically enhanced images are shown for $H_{ex} = -300$ Oe (Figure 8.5c), -600 Oe (Figure 8.5d), $H_{ex} = -900$ Oe (Figure 8.5e) and $H_{ex} = -1200$ Oe (Figure 8.5f). When the external field is increased the effective field in the sample increases: $H_{eff} = H_D + H_{ex}$. For Figure 8.5c an average effective field of about 850 Oe is obtained. Because of fluctuations of the local dipole fields and the switching fields of the individual nanowires, a few magnetic particles reverse their magnetization (black dots) also in the case $H_{eff} < H_C^{\|}$. Increasing the external field leads to an increasing number of reversed magnetized pillars (Figures 8.5b–8.5f). The enhancement of H_{eff} is partly compensated by the reduced dipole interactions from the reversed pillars. In the final image (Figure 8.5f) the applied external field has reached the coercive field $H_C^{\|} = 1200$ Oe. The number of switched (black) and unswitched (white) nanowires are nearly equal. In this case the average demagnetization field in the sample will be reduced nearly to a minimum and $H_{ex} \approx H_{eff} \approx H_C^{\|}$. Additionally our MFM images were statistically analyzed and compared with the data from the measured hysteresis loop in the preferential magnetic orientation, see Figure 8.3 ($H_{ex} \|$ wire). With a few fluctuations the MFM analysis corresponds very nicely to the bulk magnetic characterization of the sample.

In order to examine in more detail the suitability of this nickel nanowire array for patterned perpendicular magnetic media, we have tried to completely magnetized a defined area of a demagnetized sample by a strong magnetic MFM tip ($H_{tip} \approx 250$ Oe) and an external magnetic field ($H_{ex} = -1200$ Oe). The amount of the applied external field was nearly equal to the average switching field (H_{sw}) of the individual nanowire ($H_{sw} \approx H_C^{\|} = 1200$ Oe) and was applied in the direction of the nanowire axis. Starting in the upper region of Figure 8.6 the strong magnetic tip was scanned over an area of 5×5 μm². Hereby, a total external field of about $H_{ex}' = H_{ex} + H_{tip} = -1450$ Oe was

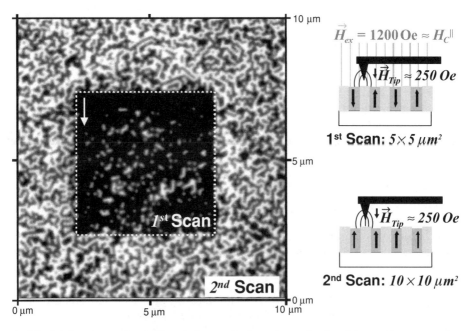

FIGURE 8.6. Local magnetic switching of a demagnetized sample area (5×5 μm², 1st scan) by a strong magnetic MFM tip ($H_{tip} \approx 250$ Oe) and an external magnetic field ($H_{ex} = -1200$ Oe). This image of the domain pattern (10×10 μm²) was recorded by a second subsequent MFM scan without an external magnetic field.

applied locally to the tips of the nickel nanowires. Subsequently, the external magnetic field was switched off. An enlarge area of $10 \times 10 \ \mu m^2$ (Figure 8.6) was scanned with the magnetic tip in order to measure the domain pattern of the manipulated area in the nanowire array. Inside the area of the first scan nearly every nickel column ($\sim$93%) is magnetized in the same direction.

Figure 8.6 shows the local impact (dark quadratic region) of the external magnetic field and the strong magnetic tip during the first MFM scan on the magnetization of nanowires. Around the magnetized region of $5 \times 5 \ \mu m$ the nickel nanowire array remained in the demagnetized state and exhibited the labyrinth-like domain pattern (Figure 8.4). The border between the magnetized area and the surrounding demagnetized area is clearly visible. From the picture, it can be concluded that the applied magnetic field ($H_{sw} \approx H_C^{\parallel}$) alone was not strong enough for the switching of magnetic polarization in the Ni columns Hence, the additional field contribution form the strong MFM tip (H_{tip}) enabled the local switching process in the Ni nanowire array.

The probability for a nickel nanowire to remain unswitched (light spots) increases in the lower region of the magnetically manipulated area (Figure 8.6, 1st scan). In the upper region, where the first magnetic scan procedure had started, the first five or six horizontal nanowire rows have been completely magnetized in the same direction. During the first scan procedure when the area of the completely magnetized nanowires was growing, the probability for nanomagnet to remain unswitched increased. We assume that the stray field interactions between the demagnetized and the magnetized area can be neglected and the net stray field in the demagnetized area is zero. By increasing the area of parallel magnetized nanowires the dipole interactions between the magnetic elements are enhanced and the applied local field ($H_{tip} + H_{ex}$) is getting less sufficient for a complete magnetic alignment of magnetization in a horizontal row of nickel columns. At the left and right border of the magnetically manipulated area the stray field interactions are weak and a lower number of unswitched magnetic columns is observed there. From this experiment, it can be concluded that the stray field dipole interactions between the nanowires are extended over several interwire distances due to their high aspect ratio (nanowire length to interwire distance: $L/D_{INT} \approx 7$). In principle, the single nanowire can store one bit of information and can be locally switched independently to the magnetization of its nearest neighbours.

8.4. NICKEL NANOWIRE ARRAYS WITH 2D SINGLE CRYSTALLINE ARRANGEMENT

By using imprint lithography as a tool for the pre-patterning of the aluminium surfaces, alumina templates with a perfect hexagonal pore arrangement on a cm^2-scale can be achieved by a single anodization process. For the first time, Masuda et al. [13] has used this technique for fabrication of perfectly ordered and unfilled alumina membranes on a small scale. In this chapter, the fabrication of Ni nanowire arrays on a cm^2-scale based on imprint lithography will be presented.

For the sample preparation, mechanically polished Al substrates were patterned by an imprint master mould described elsewhere [15]. The stamp consists of hexagonal arrays of Si$_3$N$_4$ pyramids with a pitch of $a = 500$ nm (Figure 8.7a). The imprinted etch

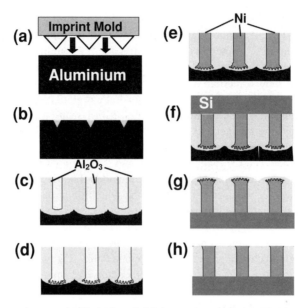

FIGURE 8.7. Preparation steps for fabrication of nickel nanowire arrays embedded in a perfectly arranged alumina matrix, which is fixed to a Si substrate. See the text for details.

pits on the Al surface act as nucleation sides for the pore formation (Figure 8.7b). The pre-structured Al surface was anodized with 1 wt.% H_3PO_4 at 195 V for 75 minutes. Alumina templates (Figure 8.7c) with a perfect hexagonal arrangement of pore channels on a cm^2-scale were obtained. Subsequently, the barrier layer was thinned at the pore bottom (Figure 8.7d) from about 250 nm down to less than 7 nm, which results in the formation of small dendrite pores at the pore bottom. Nickel was directly plated onto the nearly insulating barrier by current pulses (Figure 8.7e) and a nearly 100% pore filling was obtained. Subsequently, Si substrates were fixed on top of the area (Figure 8.7f), the Al substrate was selectively removed by chemical etching and the sample was turned upside down (Figure 8.7g). Finally, the barrier layer and the dendrite part of the nanowires were removed by etching with a focused ion beam (Figure 8.7h), in order to reduce the stray field interactions between the nanowires. Scanning electron images (Figure 8.8a) of the nanowire structure revealed $h \approx 4\,\mu m, a = 500\,nm$ and $D_P = 180\,nm$ with a dispersity $\Delta D_P / D_P < 2\%$. In comparison, Figure 8.8b shows a nickel nanowire arrays with 2D-polycrystalline arrangement of the nanowires. This nanowire array has a medium range ordering and a larger dispersity $\Delta D_P / D_P \approx 10\%$ and was fabricated by the classical two-step anodization process. Both samples were fabricated under identical electrochemical conditions.

The hysteresis loops were measured for both samples in the direction of and perpendicular to the nanowire axis (Figure 8.9). In the case when the nanowires have a monodisperse pore diameter and monocrystalline arrangement, a coercive field of 250 Oe and a remanence of 42% were detected. Because of larger dipolar interaction in the nanowire array, based on the larger deviation of the nanowire diameter and the higher disorder of the magnetic array, the second sample exhibits a reduced coercivity

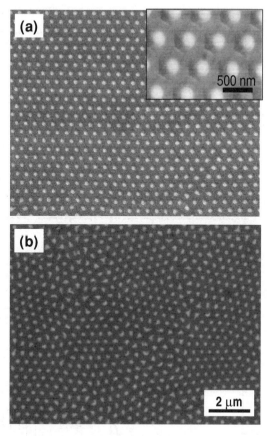

FIGURE 8.8. SEM micrographs of nickel nanowire arrays with a 2D-monocrystalline (a) and 2D-polycrystalline (b) arrangement of the magnetic columns fabricated by imprint lithography (a) and self-organization (b), respectively. Both arrays have 500 nm interwire distance and 180 nm column diameter. The length of the magnetic columns is ~5 μm. The inset in (a) shows a higher magnification of the same Ni nanowire array with the perfect arrangement.

of 160 Oe and a remanence of 30%. In contrast to our earlier results on Ni nanowire with $D_P < 55$ nm [4], a single Ni nanowire with $D_P = 180$ nm diameter does not exhibit a box-like magnetization loop. We believe that the reduced remanence of an array of nanowires is due to dipolar interactions and the sample with the 2D-monocyrstalline arrangement (Figure 8.8a) has a narrower distribution of the nanowire switching fields ($\Delta H_{sw}/H_{sw} \approx 2\Delta D_P/D_P$).

8.5. MICROMAGNETIC MODELLING

A microscopic description of the magnetic properties of small ferromagnetic particles can be obtained in the framework of micromagnetism [21,22]. The theory of micromagnetism provides the mathematical background for the calculation of magnetic structures in ferromagnets. Micromagnetism is a continuum theory in which the

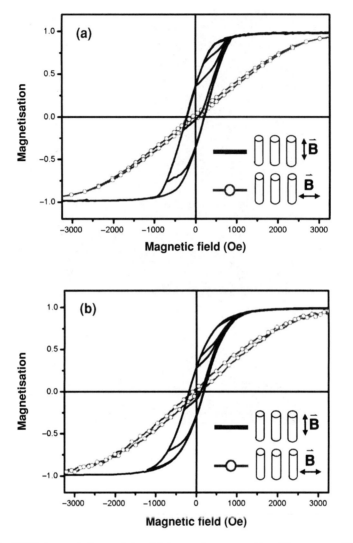

FIGURE 8.9. SQUID-hysteresis loops of the nickel nanowire arrays with a 2D-monocrystalline (a) and 2D-polycrystalline (b) arrangement measured with an applied field parallel and perpendicular to the column axis.

magnetic structure is described as a directional vector field of the magnetization $M(r)$ or the polarization $J(r) = \mu_0 M(r)$ in the sample. In principle, all the macroscopic magnetic quantities, such as the remanence, the susceptibility or the coercive field, can be obtained from this vector field if it is known as a function of space and time. Micromagnetic studies allow for a detailed analysis of the magnetic structure and magnetization processes, so that they give insight into magnetism on length scales and time scales which are experimentally difficult to access. The combination of experimental measurements and micromagnetic studies enables a deeper and broader understanding of magnetization processes and magnetic properties of nanostructures.

8.5.1. Basics of Micromagnetism

A fundamental principle of micromagnetism is that the energy of a magnet depends on the magnetic structure $M(r)$ in the sample. Usually, it is sufficient to consider the following four contributions to the micromagnetic energy density $e(r)$:

1. The exchange energy density $e_{exc} = A \times [(\nabla m_x)^2 + (\nabla m_y)^2 + (\nabla m_z)^2]$ describes the tendency of a ferromagnet to maintain a local ordering of the magnetization by aligning neighbouring magnetic moments parallel to each other. The exchange constant A is a material parameter that describes the strength of the exchange interaction, and $m = M/|M|$ is the normalized (reduced) magnetization.
2. The stray field energy density $e_{stray} = -J \cdot H_d/2$ is connected with the magnetostatic interaction between the magnetic moments. The stray field H_d is the field that results from the dipolar fields of all the magnetic moments in the sample.
3. The Zeeman term $e_{ext} = -J \cdot H_{ext}$ describes the influence of an external field. The magnetization tends to align parallel to the applied field.
4. In crystalline materials the energy of the magnetic structure depends on the direction of the magnetization with respect to the crystalline axes. In the simplest form, the magnetocrystalline anisotropy energy density is $e_{ani} = K_u \sin^2\alpha$, where K_u is the uniaxial anisotropy constant and α is the angle enclosed between the magnetization and the easy axis.

Static magnetic structures represent a minimum of the total energy of the sample. This is not necessarily a global minimum. In numerical calculations equilibrium magnetization structures can be obtained by minimizing the total energy of the magnet with respect to the discretized directional field of the magnetization.

The magnetization dynamics is governed by the Landau–Lifshitz–Gilbert equation

$$dM/dt = -\gamma M \times H_{eff} + \alpha/M_s(M \times dM/dt) \qquad (8.1)$$

where γ is the gyromagnetic ratio and α is a phenomenological damping constant [23]. The equation describes a combined precession and relaxation motion of the magnetization in an effective field H_{eff}. The effective field contains contributions from the aforementioned energy terms. It is obtained from the local energy density by means of a variational derivative with respect to the magnetization $\mu_0 H_{eff} = -M_s^{-1}\delta e/\delta m$. In dynamic micromagnetic simulations, the Landau–Lifshitz–Gilbert equation is used to calculate the evolution of the magnetization $M = M(r, t)$ in time and space. The temporal evolution of M is important if the magnet is not in an equilibrium. Such a non-equilibrium situation is given, e.g., during the magnetization reversal process in an external field.

In soft-magnetic materials, the exchange constant $\lambda = (2A/\mu_0 M_s^2)^{1/2}$ is an intrinsic material-dependent length scale that (roughly) describes the typical spatial extension of inhomogeneities in the magnetic structure, such as, e.g., domain walls.

8.5.2. Computational Micromagnetism with the Finite Element Method

Generally, analytic solutions of micromagnetic problems are only possible if strong simplifications are assumed. Owing to the tremendous progress in computer speed as well as in numerical techniques in the last years, it has become possible to reliably calculate both static magnetic structures and dynamic magnetization processes in nanoscaled ferromagnetic particles. There are several numerical difficulties involved in micromagnetic simulations, including the accurate and fast calculation of the long-range magnetostatic interaction given by the stray field and the stable integration of the Landau–Lifshitz–Gilbert equation. Over the last years, several techniques have been developed to treat these problems. The finite element method is a particularly powerful tool when micromagnetic problems of particles with curved boundaries are to be solved. The versatility of the finite element method is due to the geometrical flexibility connected with the discretization cells. By using cells of irregular tetrahedral shape, the particle's shape can be approximated particularly well, in contrast to the more frequently used finite-difference schemes, where simulating particles with curved boundaries is problematic [24].

The calculation of the stray field is performed by introducing a scalar magnetic potential U. The numerical solution of Poisson's equation $\Delta U = \nabla M$ with accurate consideration of the boundary conditions is obtained using a combination of the finite element method with the boundary element method (FEM/BEM). The stray field is derived from U as a gradient field $H_d = -\nabla U$. A detailed description of this precise and fast method is given elsewhere [25]. A particularly advantageous feature of the FEM/BEM scheme is the possibility of calculating the magnetostatic interaction of separate magnetic particles by simply placing the finite element meshes next to each other, without the need to perform time-consuming calculations of the field in the space between the particles. In the present case, this allows us to simulate the influence of magnetostatic coupling in an array of nanowires.

8.5.3. Magnetostatically Coupled Nickel Nanowires

The magnetic properties of sets of hexagonally ordered nickel nanowires have been simulated by means of micromagnetic finite element modelling [26]. The model has been chosen according to the experimental situation described elsewhere [18]. Several nanowires are placed on a hexagonal array with 100 nm period. The wires' diameter is $d = 40$ nm, and their length is $l = 1$ µm. The saturation polarization is $J_s = 0.52$ T and the exchange constant is $A = 10.5$ pJ/m. The wires are assumed to be amorphous, hence the magnetocrystalline anisotropy is zero, $K_u = 0$ J/m^3.

The influence of the magnetostatic interaction on the coercive field of an array of nickel nanowires is illustrated in Figure 8.10b. By placing an increasing number of wires on hexagonal lattice sites, the coercive field is calculated for different numbers of interacting wires. The calculation is performed statically by means of energy minimization. The simulations yield a significant decrease of the coercive field from 145 mT in the case of a single, isolated wire to 115 mT in the case of 16 interacting nanowires. A fully micromagnetic simulation without simplifications can hardly be performed on considerably larger arrays because of the numerical costs. However, the tendency shown in Figure 8.7b is clear: The coercive field of the array is strongly reduced by the magnetostatic interaction. Compared to the small number N of wires considered in the simulation, the experimental situation corresponds rather to the case $N \to \infty$. Although the number of points N is not sufficient here for a convincing extrapolation to infinity, it is evident that the experimentally observed value of the coercive field $\mu_0 H_c \approx 100$ mT can only be

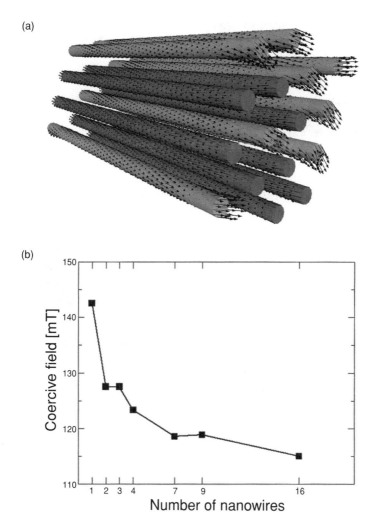

FIGURE 8.10. (a) Micromagnetic simulation of 16 interacting nanowires at the coercive field. Half of the wires have switched in the direction of the field (dark grey). The wires are magnetized homogeneously along the axis. (b) Coercive field of small hexagonal arrays of nanowires as a function of the number of interacting wires.

explained as a consequence of the magnetostatic coupling between the wires. Concerning the magnetic structure of the individual wires, the simulations indicate that the idealized case, according to which the wires are magnetized homogeneously along the wire axis, is fulfilled almost exactly. In an easy-axis hysteresis loop (field applied parallel to the wire axis) the magnetization of each wire is either parallel or anti-parallel to the field. The pronounced magnetostatic shape anisotropy of the wires enforces an alignment of the magnetization parallel to the axis. An example for this is given in Figure 8.10a, where the magnetic structure of a small array consisting of 16 nanowires is displayed. After saturation, the array has been exposed to a reversed field close the coercivity $\mu_0 H_c \approx 115$ mT.

Some wires (dark grey) have switched towards the field; others (light grey) are still aligned anti-parallel to the field. Each wire is a magnetically bi-stable particle.

8.5.4. Magnetization Reversal Dynamics in Nickel Nanowires

To calculate the switching speed and to study the reversal dynamics, micromagnetic simulations based on the Landau–Lifshitz–Gilbert equation have been performed on single nanowires [27]. In all cases, the wires have been exposed instantaneously to a 200 mT field, which is sufficiently strong to revert the magnetization. A Gilbert damping constant $\alpha = 0.1$ is assumed. The dynamic magnetization reversal process of a nickel nanowire of the set discussed above is shown in Figure 8.11a. The reversal begins at the wire's ends. This is where the demagnetizing field has its strongest value.

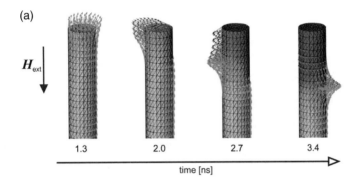

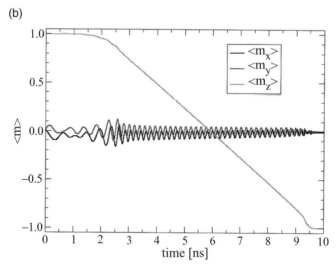

FIGURE 8.11. (a) Snapshots of the initial stages of the magnetization reversal of a Ni nanowire ($d = 40$ nm) via the transverse wall mode. The magnetization reverts in a nucleation-propagation process that starts at the wire's ends. The transverse component of the 180° head-to-head wall precesses in the external field and leads to a spiralling motion of the wall. (b) Average magnetization components along the axis (m_z) and perpendicular to it (m_x, m_y) as a function of time during the reversal.

The demagnetizing field adds to the externally applied field, thus facilitating the magnetization reversal. The nucleated domain, in which the magnetization has switched, is separated from the non-reversed part of the wire by a 180° head-to-head wall [28,32,33]. As the domain wall propagates along the wire axis, the reversed domain expands until the entire sample is switched. In the middle of the domain wall, the magnetization points perpendicular to the wire axis. Since the field is applied along the wire axis, a strong torque is exerted on the magnetization in this region of the domain wall. This leads to a precessional motion of the magnetization in the domain wall, which proceeds on a characteristic spiralling orbit along the wire axis. This motion reflects in oscillations of the magnetization components perpendicular to the wire axis, see Figure 8.11b.

The magnetic structure in this wire is one-dimensional during the reversal, i.e., the direction of the magnetization depends only on the position along the wire axis. It is homogeneous on any cross-section through the wire. Obviously, this case is only realized in very thin wires. As the wire thickness increases, the magnetic structure that is formed during the reversal process becomes three-dimensional. An example for a three-dimensional reversal mode is shown in Figure 8.12a, where the initial stages of a switching process in a slightly thicker nickel nanowire (diameter: 60 nm) are displayed.

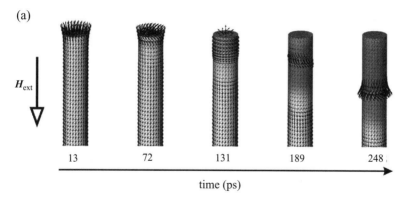

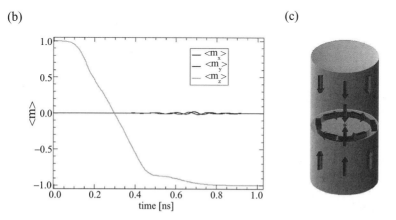

FIGURE 8.12. (a) Magnetization reversal in a Ni nanowire ($d = 60$ nm) via the vortex mode. This reversal process is considerably faster than the transverse mode (b). The switching is accomplished by the nucleation and subsequent propagation of an axial vortex wall. (c) Schematic representation of the vortex wall with a singularity in the middle.

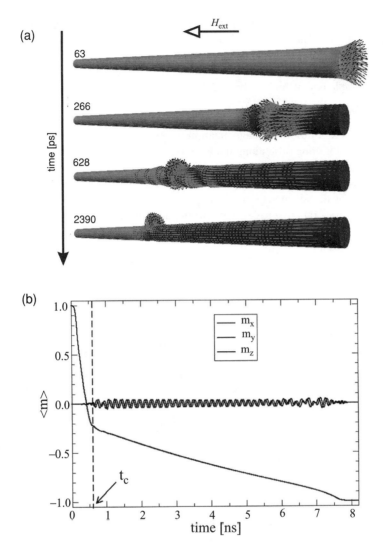

FIGURE 8.13. (a) Dynamic mode conversion in a cone-shaped wire of 1 μm length and linearly varying diameter between 30 and 60 nm. The reversal starts at the thicker end as a vortex mode. The reversal front propagates along the wire. When it passes through a range of critical thickness, the vortex mode converts into the corkscrew mode. (b) Average magnetization components during the reversal of the cone-shaped particle. The mode conversion sets in at the time t_c. At that point, the slope of m_z changes drastically, indicating the conversion of the fast vortex reversal mode into the slow corkscrew mode. The conversion into the corkscrew mode also reflects in the onset of the characteristic oscillations in the m_x and m_y components.

The reversal mechanism is again a nucleation-propagation process that starts at the wire's ends. But now the reversed region is separated from the non-reversed region by a vortex wall. The structure of a vortex wall is sketched in Figure 8.12c. It is interesting to note that this structure contains a micromagnetic singularity, known as Bloch point [17,29]. This vortex reversal process has some similarities with the classical curling reversal mode [21,30]. The main difference is the localization of the mode [20]. This reversal process is considerably faster than the aforementioned mode with a propagating transverse domain wall, cf. Figures 8.11b and 8.12b. Hence, the wires should not be too thin if they are to be switched quickly.

The thickness-dependent transition from the vortex wall reversal mode to the transverse wall mode can be nicely simulated in a cone-shaped wire [31]. Figure 8.13a shows an example of the magnetization reversal in a Ni wire with linearly varying thickness ($d = 30$ nm on one end and $d = 60$ nm on the opposite end). The reversal begins with the vortex mode at the thicker end. As the vortex wall proceeds along the wire axis, it passes through regions of constantly decreasing diameter. When the reversal front reaches a region of critical thickness (here $t_c \approx 42$ nm), a spontaneous mode conversion occurs. The mode conversion is clearly visible when the average magnetization components are plotted as a function of time, as shown in Figure 8.13b. At the beginning, the reversal proceeds quickly via the vortex reversal mode (steep slope of the m_z component, $z =$ wire axis). After the conversion into the transverse wall mode, the m_z-slope changes abruptly, and the reversal speed is strongly reduced. In this case, about 90% of the switching time is required for the last 30% of the reversal. Besides the change in reversal speed, another effect connected with the mode conversion is the onset of the oscillations in the m_x and m_y components with the beginning of the transverse wall reversal mode.

8.6. CONCLUSION

The measurement of the bulk-magnetic properties shows a strong magnetic anisotropy along the nickel column axes, coercive fields of 1200 Oe and nearly 100% squareness. In the demagnetized state the nanowire array exhibits a labyrinth-like domain pattern. Good agreement between the MFM investigation in the presence of an external magnetic field and the hysteresis loop was obtained. Each magnetic pillar is a single-domain magnetic particle, magnetized perpendicular to the template surface and, in principle, can store one bit of information. By using imprint lithography perfectly arranged nickel nanowire arrays have been fabricated. The properties of the magnetic arrays depend on the ordering degree of the nanowire arrangement. If the deviation of the nanowire diameter decreases and the ordering degree of the nanowires array is enhanced, the anisotropy of the whole nanowire array increases. Micromagnetic simulations show that the magnetostatic interaction between the wires has a decisive influence on the coercive field of a nanowire array. Good agreement with the experiment is obtained by simulating an array of 16 interacting wires. Simulations of the magnetization reversal dynamics in single nanowires predict that the magnetization can switch via two different nucleation-propagation modes. Which of the mode occurs depends on the wire thickness. The transverse mode that occurs in thinner wires ($d \approx 30$ nm) leads to a considerably slower reversal than the vortex mode found in thicker wires ($d \approx 60$ nm).

ACKNOWLDGEMENTS

We are grateful to T. Schweinböck, H. Kronmüller, S. F. Fischer, D. Weiss, D. Navas and M. Vazquez for support, experimental measurements or fruitful discussions. We also like to thank Professor U. Gösele and Professor J. Kirschner for supporting this activity.

REFERENCES

[1] R. O'Barr, S.Y. Yamamoto, S. Schultz, W. Xu and A. Scherer, J. Appl. Phys. **81**, 4730 (1997).

[2] C.A. Ross, H.I. Smith, T.A. Savas, M. Schattenberg, M. Farhoud, M. Hwang, M. Walsh, M.C. Abraham and R.J. Ram, J. Vac. Sci. Technol. B **17**, 3159 (1999).

[3] D. Routkevitch, A.A. Tager, J. Haruyama, D. Almawlawi, M. Moskovits and J.M. Xu, IEEE Trans. Electron Devices **147**, 1646 (1996).

[4] M.P. Kaneko, IEEE Trans. Magn. **17**, 1468 (1981).

[5] D. Al Mawlawi, N. Coombs and M. Moskovits, J. Appl. Phys. **69**, 5150 (1991).

[6] L. Feiyue, R.M. Metzger and W.D. Doyle, IEEE Trans. Magn. **33**, 3715 (1997).

[7] G.J. Strijkers, J.H.J. Dalderop, M.A.A. Broeksteeg, H.J.M. Swagten and W.J.M. de Jonge, J. Appl. Phys. **86**, 5141 (1999).

[8] H. Zeng, M. Zheng, R. Skomski, D.J. Sellmyer, Y. Liu, L. Menon and S. Bandyopadhyay, J. Appl. Phys. **87**, 4718 (2000).

[9] H. Masuda and K. Fukuda, Science **268**, 1466 (1995).

[10] A.-P. Li, F. Müller, A. Birner, K. Nielsch and U. Gösele, J. Appl. Phys. **84**, 6023 (1998).

[11] K. Nielsch, J. Choi, K. Schwirn, R.B. Wehrspohn and U. Gösele, Nano Lett. **2**, 677 (2002).

[12] A.P. Li, F. Müller and U. Gösele, Electrochem. Soc. Lett. **3**, 131 (2000).

[13] H. Masuda, H. Yamada, M. Satoh, H. Asoh, M. Nakao and T. Tamamura, Appl. Phys. Lett. **71**, 2770 (1997).

[14] K. Nielsch, F. Müller, A.P. Li and U. Gösele, Adv. Mater. **12**, 582 (2000).

[15] J. Choi, K. Nielsch, M. Reiche, R.B. Wehrspohn and U. Gösele, J. Vac. Sci. Technol.B **21**, 763 (2003).

[16] K. Nielsch, F. Müller, R.B. Wehrspohn, U. Gösele, S.F. Fischer and H. Kronmüller, *The Electrochemical Society Proceedings Series*, PV 2000-8, Pennington, NJ, 2000, p. 13.

[17] E. Feldtkeller, Z. Angew. Phys. **19**, 530 (1965).

[18] K. Nielsch, R. Wehrspohn, J. Barthel, J. Kirschner, U. Gösele, S.F. Fischer and H. Kronmüller, Appl. Phys. Lett. **79**, 1360 (2001).

[19] K. Ounadjela, R. Ferré, L. Louail, J.M. George, J.L. Maurice, L. Piraux and S. Dubois, J. Appl. Phys. **81**, 5455 (1997).

[20] H.B. Braun, J. Appl. Phys. **85**, 6172 (1999).

[21] A. Aharoni, *Introduction to the Theory of Ferromagnetism*, Oxford Science Publications, Clarendon Press, Oxford, 1996.

[22] W.F. Brown, Jr., *Micromagnetics*, Interscience Publishers, John Wiley & Sons, New York, London, 1963.

[23] T.L. Gilbert, Phys. Rev. **100**, 1243 (1955).

[24] C.J. Garcia-Cervera, Z. Gimbutas and Weinan E, J. Comp. Phys. **184**, 37 (2003).

[25] D.R. Fredkin and T.R. Koehler, IEEE Trans. Magn. **26**, 415 (1990).

[26] R. Hertel, J. Appl. Phys. **90**, 5752 (2001).

[27] R. Hertel, J. Magn. Magn. Mater. **249**, 251 (2002).

[28] R.D. McMichael and M.J. Donahue, IEEE Trans. Magn. **33**, 4167 (1997).

[29] R. Hertel and H. Kronmüller, J. Magn. Magn. Mater. **238**, 185 (2002).

[30] E.H. Frei, S. Shtrikman and D. Treves, Phys. Rev. **106**, 446 (1957).

[31] R. Hertel and J. Kirschner, Physica B **343** (2004) 206.

[32] D. Hinzke and U. Nowak, J. Magn. Magn. Mater. **221**, 365 (2000).

[33] H. Forster, T. Schrefl, D. Suess, W. Scholz, V. Tsiantos and J. Fidler, J. Appl. Phys. **91**(10), 6914 (2002).

9

Porous Silicon for Micromachining

P.J. French[1] and H. Ohji[2]

[1]*Electronic Instrumentation Laboratory, Department of Microelectronics, Faculty of Electrical Engineering, Mathematics and Computer Science, Delf University of Technology, Mekelweg 4, 2628 CD Delf, The Netherlands*
[2]*Mitsubishi Electric Corporation, Advanced Technology Research and Development Centre, Amagasaki, Hyogo 6618661, Japan*

9.1. INTRODUCTION

The basic porous silicon processing has been used for micromachining for many years. In the early days, this was usually in the form of electropolishing. This was found to be an effective technique to etch through the full thickness of the wafer leaving a thin membrane. For example, etching through an n+ wafer and stopping on an n-type membrane has been applied [1]. This had the advantage of producing a well-defined membrane thickness with a relatively simple process. If the current density was too low and/or the HF concentration too high, porous silicon was formed on the surface. In these early days, this was simply seen as an etching experiment gone wrong. In the 1990s, the potential for micromachining was seen [2–4]. The material was found to be an effective technique for forming SOI and micromachining. The porous silicon could be oxidized at low temperature or was able to hold the structures in position to be removed at a later stage. Due to the extremely high surface area, the porous silicon can be removed very easily, for example in KOH at room temperature or even developer for photoresist. Furthermore, since the etching is an electrochemical process the porous formation can be performed selectively. The above processes all used microporous silicon. In recent years, macroporous silicon has also been shown to be highly effective for micromachining. Macroporous silicon was first used to make deep straight holes for capacitors [5,6]. The same process has since been shown to be effective for the fabrication of truly 3D mechanical structures [7].

This chapter gives an overview of the development of both micro and macroporous silicon techniques for micromachining. A comparison is given between the different techniques with a number of devices to illustrate the potential.

9.2. BASIC PROCESS

Porous silicon formation is generally an electrochemical process, although purely chemical etching recipes based on HF/HNO3 exist to create porous silicon. In addition to HF, a number of alternative etchants can be used. Some examples of available etchants are as follows:

- HF,
- HF + surfactants,
- AFEM,
- HF-dimethyl-formamide.

The basic process, which requires two electronic holes to remove one silicon atom from the surface of the silicon, is the following two-step process:

$$Si + 2h^+ + 2H^+ \rightarrow Si^{4+} + H_2 \tag{9.1}$$

$$Si^{4+} + 6HF \rightarrow SiF_6^{2-} + 6H^+. \tag{9.2}$$

Once an atom has been removed the chance of further removal from that point is greater than the surrounding area. Thus, the formation of porous silicon is able to continue. This also shows the reason why p-type silicon is more readily made porous. For n-type silicon, illumination is usually required to generate the holes. The basic process set-up is given in Figure 9.1.

The basic process is based on HF solutions. In the first solution, only HF is used. The addition of a surfactant is used to reduce surface tension and also ensure that the H_2 bubbles formed have the chance to leave the surface before they get too large [8]. The next etch solution is ammonium fluoride etch mixture (AFEM) [9]. This solution become of interest since it is able to form porous silicon but has a low etch rate for aluminium, which, since porous formation is often used as a post-processing step, is often present. The potential problem with AFEM is that there appears to be an additional chemical component in the etching that may result in unwanted etching. However, with some additional care this etchant can be used. The interest in the fourth etchant, DMF, began with the desire to make macroporous silicon in p-type material [10]. Macroporous silicon has a quite different structure from the microporous material. Micropores have a random structure whereas macroporous are regular [11]. Since that time it has also been shown that DMF can also be used to make porous silicon in n-type silicon [12].

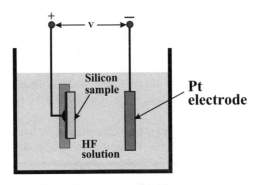

FIGURE 9.1. Basic process set-up for porous formation.

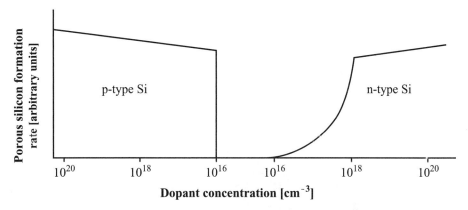

FIGURE 9.2. Porous silicon formation rate for n- and p-type silicon.

This etchant has a low etch rate for oxide, but a reasonable porous formation rate. This is discussed in more detail later.

Figure 9.2 shows that p-type silicon is most readily made porous due to the fact that electronic holes are required in the process [2]. As is discussed later, n-type porous formation requires illumination to generate electronic holes. Therefore, in the non-illuminated condition regions in the wafer can be selectively made porous. This is usually in the form of making p-type selectively porous. These are all microporous structures.

9.2.1. Micromachining Using Microporous Silicon

The basic structure of the microporous silicon is shown in Figure 9.3 [11]. This shows a sponge-type structure. This type of structure was first observed by Canham in 1990 to emit visible light [13]. This first observed emission was in the form of photoluminescence, but subsequently both photo- and electroluminescence have been observed [14–17]. Other applications for microporous silicon include humidity sensors [18], gas sensors [19] and thermal isolation [20], where the porous silicon can easily be oxidized without mechanical stress through expansion. Its potential as a sacrificial layer was presented in the early 1990s [2–4]. The advantages of this material are that it can

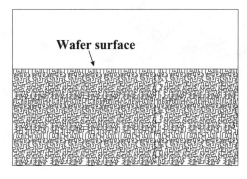

FIGURE 9.3. Structure of p type porous silicon.

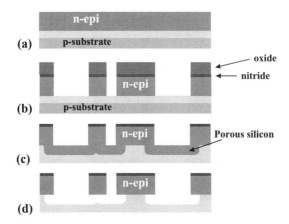

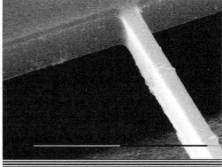

FIGURE 9.4. Basic process sequence for sacrificial porous silicon.

hold the mechanical structures in position until the end of the processing and then can be removed very easily due to the large surface area. For example, KOH at room temperature or resist developer can be used. For a review on porous silicon as a sacrificial layer, see [21].

As described above, the preference for p-type formation and the fact that it is an electrochemical process is extremely useful for micromachining. The basic process sequence is shown in Figure 9.4 [22]. Free-standing epi structures can be fabricated by making the substrate porous and then removing the porous layer at a later stage. The advantage is that the n-type epi is not attacked by porous formation and therefore does not need sidewall protection. The basic process is thus as follows: the starting point is the n-type epi on the p-type substrate (Figure 9.4a). Using an oxide, or other mask, the epi is defined to reveal the substrate (Figure 9.4b). Then using a back contact, in the configuration shown in Figure 9.1, the exposed regions of the substrate are made porous (Figure 9.4c). Finally, the porous silicon is removed to create free-standing structures (Figure 9.4d).

An example of a fabricated structure is given in Figure 9.5. The picture on the right shows the nitride mask on the top of the epi and the air gap under the epi.

FIGURE 9.5. Free-standing structures fabricated using sacrificial porous formation.

vert acc (epi etched away)   10μm

FIGURE 9.6. Partially released structure showing the wine-glass effect.

Figure 9.5 (left) shows the characteristic ridges between the etch holes. This is due to reduced current as the un-etched region narrows, and is often referred to as the wine-glass effect. This effect can be seen more clearly in Figure 9.6.

If aluminium is already on the frontside of the wafer this should be protected or an etchant should be used which does not attack it. As mentioned above, AFEM is a suitable option and such an example is given in Figure 9.7 [9]. However, this does present

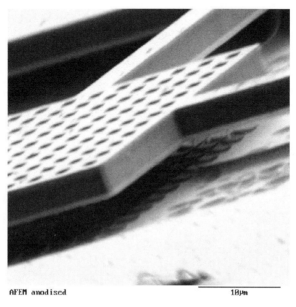

AFEM anodised   10μm

FIGURE 9.7. Micromachined structure fabricated using AFEM.

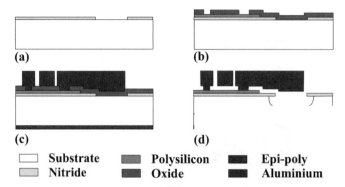

(a)   (b)

(c)   (d)

☐ Substrate   ▨ Polysilicon   ▧ Epi-poly
☐ Nitride   ▨ Oxide   ▧ Aluminium

FIGURE 9.8. Basic process of epi poly combined with porous formation.

some other problems since this introduces a chemical element to the etching and it is no longer purely electrochemical. In order to protect the n-type epi, which would have been etched chemically by the etchant, a silicon nitride layer is needed after the definition of the structures. The lighter colour at the base of the structure is where the epi has been etched leaving only the thin nitride layer.

Porous formation can also be used to increase the air gap under surface micromachined or epitaxial structures. When oxide is used as a sacrificial layer a single etch step can be used. First, the oxide is removed using chemical etching and then porous formation is begun by applying a voltage.

This technique has also been combined with epi-poly processing [23]. In a similar process epi-poly structures, 4–8 μm thick, were produced with large air gap underneath (Figures 9.8 and 9.9).

FIGURE 9.9. SEM photograph of structure fabricated using the double sacrificial etch.

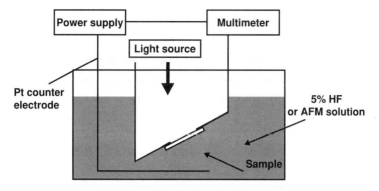

FIGURE 9.10. Etch set-up for the macroporous formation in n-type silicon.

9.2.2. Micromachining with Macroporous Silicon

The macroporous process was first proposed by Lehmann for making deep straight holes with high aspect ratios [5]. The basic process is as follows. As shown above, porous formation requires electronic holes, which are in short supply in n-type silicon. However, illuminating the backside of the sample will generate electron–hole pairs (Figure 9.10). Alternatively, a liquid contact can be used. The holes move to the front of the wafer and start the etching process. This is illustrated in Figure 9.11.

This process was developed for making large capacitors [6]. The basic process is given in Figure 9.12. The starting material is an n-type wafer (Figure 9.12a). On this wafer, a SiN making layer is deposited and patterned to define the macropores (Figure 12b). A starting point for the pores is made using KOH (Figure 9.12c) and then electrochemically etched in HF to form straight pores (Figure 9.12d). Using this technique high aspect ratios can also be achieved, as shown in Figure 9.13.

A further interesting feature of this process can be seen in Figure 9.14. This shows that the current density does not control the vertical etch rate, but the width of the pores.

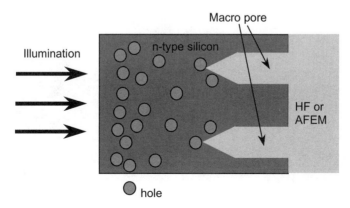

FIGURE 9.11. Etch process in n-type silicon.

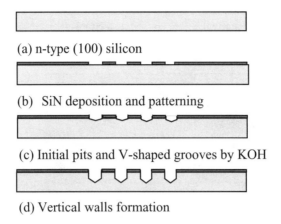

(a) n-type (100) silicon

(b) SiN deposition and patterning

(c) Initial pits and V-shaped grooves by KOH

(d) Vertical walls formation

FIGURE 9.12. Basic process sequence for macroporous formation.

The process can therefore be modified to make free-standing structures. This is shown in Figure 9.15. The first three steps are the same as the standard process. Then, when the required depth is reached, the light intensity is increased to increase the pore width and release the structures (Figure 9.15d).

Therefore three-dimensional structures can be manufactured, since the change in width only occurs at the tip of the pore. An example of a fabricated structure is given in Figure 9.16.

The limitation of this process is the use of KOH to form the starting point of the pore. This means that the holes or slots have to be orientated in the $\langle 110 \rangle$ direction as (100) wafers are used. The initial assumption was that the point formed by the KOH was necessary to start the pore. However, it has been shown that plasma etching or isotropic etching can be used [24]. This removes the limitation of KOH and thus any shape can be fabricated. Examples of the structures can be seen in Figure 9.17.

The above discussion has been limited to n-type silicon because of the abundance of holes in p-type silicon.

FIGURE 9.13. Deep straight trenches fabricated using macroporous techniques [6]. Reproduced with kind permission from Fred Roozeboom, Philips, The Netherlands.

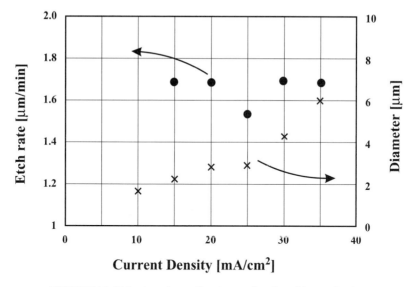

FIGURE 9.14. Etch rate and pore diameter as a function of current density.

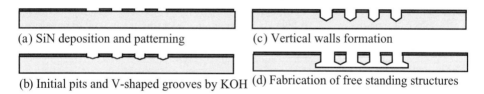

(a) SiN deposition and patterning   (c) Vertical walls formation

(b) Initial pits and V-shaped grooves by KOH (d) Fabrication of free standing structures

FIGURE 9.15. Adjusted macroporous silicon process for 3D structures.

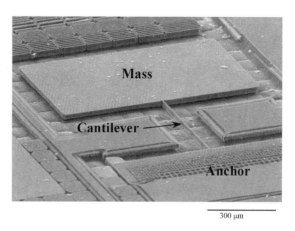

300 μm

FIGURE 9.16. Free-standing structure fabricated using macroporous silicon micromachining [7].

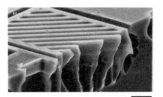

10 µm   10 µm

(a) Initial grooves made by RIE   (b) Trench structures

FIGURE 9.17. Structures fabricated using RIE starting point.

Macroporous silicon in p-type silicon is more difficult than n-type. Initially, using HF, only micropores were formed on p-type silicon substrate und thus trenches could not be achieved. Recently, in order to form macropores in p-type silicon, a new etchant, DMF, has been proposed [10]. The basic constituents of this etchant are: 4% HF (50%), 8% tetrabutylammonium perchlorate (TBAP), 4% H_2O and the rest was dimethyl-formamide [25,26]. Being p-type silicon, with an abundance of holes, no light is required to achieve etching. However, it has also been shown that accurate control of the pore width cannot be achieved through current density, although micromachining can be performed using a switch between porous formation and electropolishing. SEM micrographs of etched surface as a function of current density for 20 minutes etch time are given in Figures 9.18a–c. In the case of lower current density, the whole area has been etched down uniformly and thus the initial grooves can be seen after electrochemical etching (Figure 9.18a). On the other hand, smooth etched surface was obtained due to electropolishing as shown in Figure 9.18c. Trench structures can be seen with the current density of 20 mA/cm^2 in Figure 9.18b although the etched surface was slightly rough. A close-up view of the trench structures is given in Figure 9.19. In order to fabricate free-standing beams, the current density increased after making the trench structures. For example, a current density of 20 mA/cm^2 can be used to obtain trenches followed by a current density of 30 mA/cm^2 to connect the trenches [10]. Free-standing beams can be achieved with this etching technique as shown in Figure 9.19. Clear gap between the beam and the substrate can be seen.

The DMF etchant is more difficult to control but in some cases it may be an advantage that the etch rate of oxide is low. The etch rate of oxide in HF and DMF is shown in Table 9.1. Thermally grown oxide can be seen to be a suitable mask with this etchant.

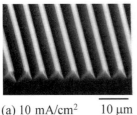
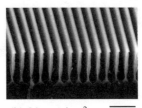
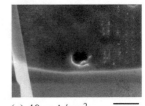

(a) 10 mA/cm^2 10 µm   (b) 20 mA/cm^2 10 µm   (c) 40 mA/cm^2 10 µm

FIGURE 9.18. Etch results for three currents.

TABLE 9.1. Comparision of H_2O-based etchant with DMF-based etchant with etch rates of various masking materials. Both etchant contains 5 wt % HF. All etch rates are in $\overset{\circ}{A}$/min.

	H_2O-based HF (5 wt.%) etchant	DMF-based HF (5 wt.%) etchant
Thermally grown SiO_2	260	8
PECVD-TEOS	220	7
PECVD-SiN	95	6
Sputtered SiN	39	1.4
LPCVD-SiN	12	1.4

Further studies have been performed to use DMF for an n-type material. As in the case of HF, light is used to control the current. However, unlike HF, this cannot be used to control the pore diameter, as shown in Figure 9.20.

10 µm

FIGURE 9.19. Close-up view of micromachined beams showing the air gap.

One of the problems with this etchant is its low conductivity. Therefore, the current, which is searching for the lowest resistive path is more likely to travel through the silicon between pores causing side branching. Tetrabutylammonium perchlorate (TBAP) powder was added to this etchant to control conductivity of etchant. As shown in Figure 9.21, increasing levels of TBAP leads to increased conductivity and smoother walls.

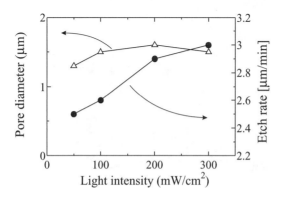

FIGURE 9.20. Effects of light intensity on pore diameter and etch rate. Applied voltage was set at +1.2 V.

	Without TBAP	10g TBAP	20g TBAP	30g TBAP
Cross-section view of etched pore (Magnification is 3500)	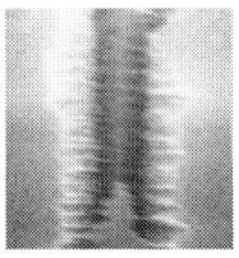 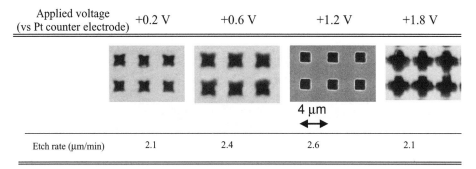	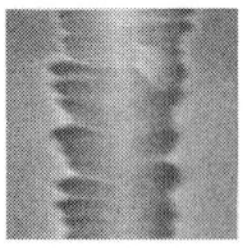	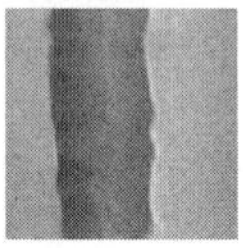	
Conductivity of etchant	4 mS/cm	22 mS/cm	29 mS/cm	33 mS/cm
Etch rate of pore	0.4 µm/min	1.8 µm/min	2.2 µm/min	2.6 µm/min

FIGURE 9.21. Effect of TBAP on the conductivity of etchant and the quality of pores.

Applied voltage (vs Pt counter electrode)	+0.2 V	+0.6 V	+1.2 V	+1.8 V
	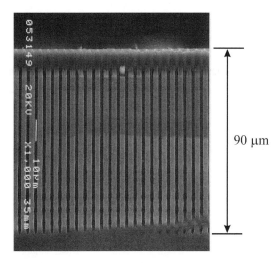		4 µm ↔	
Etch rate (µm/min)	2.1	2.4	2.6	2.1

FIGURE 9.22. Plain view of pore morphology and etch rate influenced by the applied voltage, for etching n-type material. The light intensity was 100 mW/cm^2. The etchant composition was 270 g DMF, 15 g HF, 15 g H$_2$O, and 30 g TBAP. The etching was performed for 30 minutes.

90 µm

FIGURE 9.23. Cross-sectional view of pores, etched, using: light intensity 300 mW/cm^2, applied voltage 1.2 V; etchant: DMF 270 g, HF 15 g, H$_2$O 15 g and TBAP 30 g.

There also appears to be some crystal dependence which can change with applied voltage. Etched samples were mechanically polished for a depth of 10 μm to observe the plain view of etched sample. At applied voltage of +0.2 V, pore shape is like star shape that has (110) preferred crystal orientation. At higher voltage of +1.8 V, pore shape becomes the cross shape that has (100) preferred crystal orientation. The pore wall exhibits a preferred crystal orientation, which is strongly affected by applied voltage. This is illustrated in Figure 9.22.

Electropolishing, in which etched surface becomes smooth, is not observed in DMF-based HF etchant, which is different from H_2O-based HF etchant. By optimizing the etching parameters, high aspect ratio pores can be achieved. This is illustrated in Figure 9.23.

9.3. APPLICATIONS

9.3.1. Sealed Cavity Devices

One important difference in the sacrificial porous silicon technique compared to wet chemical etching is that the pores grow from the tips of the pores and therefore changes in the etch parameters will only affect the new porous regions and not those already formed. This feature has been used to form cavities by Anderson *et al.* [27]. In this case, the mechanical layer used was silicon nitride and the sacrificial layer was polysilicon. The resulting structure is shown in Figure 9.24. To fabricate this structure three steps were used. The first step, using 5% HF and 5 V dynamic hydrogen electrode (DHE), removed the first part of the cavity. This was followed by a high HF concentration step (49% HF) at 0 V DHE, which formed the porous plug. Finally, 10% HF at 5 V DHE was used to form the inner cavity.

9.3.2. Free-Standing Porous Structures

In most micromachining processes, the porous silicon is used as a sacrificial layer. However, it can also be used as a mechanical structure. In Figure 9.23, an example regarding this is given [28]. In the first part of the process, first a porous layer is made in the substrate and once the desired depth is reached the current is increased to switch to electropolishing (Figure 9.25a). This structure is held in place due to the porous formation under the mask. Removal of the mask then releases the porous structure. This

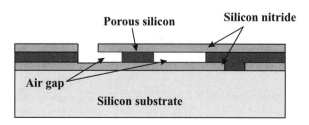

FIGURE 9.24. Cross-section of the sealed cavity fabricated using porous the formation of polysilicon [27].

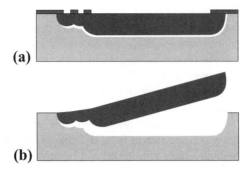

FIGURE 9.25. Process sequence: (a) after double porous formation process and (b) after full release [28].

release is performed using dry etching in an oxygen plasma. This has the advantage of avoiding stiction. It also causes a partial oxidation of the pores and through this produces a compressive stress. This is the reason for the lifting up of the mechanical structures (Figure 9.25b).

9.3.3. Isolated Structures

Polysilicon bridges constructed using porous silicon as a sacrificial layer have been fabricated for various sensor applications. Porous silicon formation and removal release the bridges, forming flow channels beneath them (Figure 9.26). Bridges constructed of thin films have small cross sections that allow small time constants and very high speed sensing. This type of structure can be used in fabricating devices such as flow sensors, vacuum sensors, hot-wire anemometers and gas sensors [29].

A similar structure was presented for a flow sensor using a broader bridge enabling the integration of thermopiles. In this case, the flow is over the top surface of the chip [30].

A thin-film bolometer, consisting of a thin metal resistor on the top of a silicon carbide membrane, has also been fabricated using porous silicon as the sacrificial material [4]. The large separation between the membrane and the substrate that is possible using

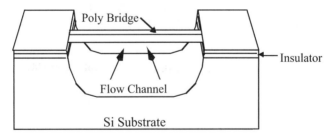

FIGURE 9.26. An example of a polysilicon bridge that could be used as a flow or other type of sensor. The flow channel beneath the bridge is formed by making part of the substrate porous, then removing the porous material in a hydroxide solution. Taken from [4].

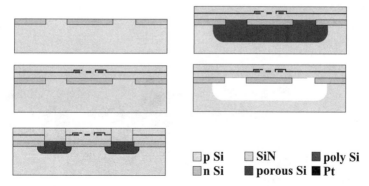

FIGURE 9.27. Process sequence of the bulk micromachined micro hotplate. Redrawn from [31].

this technique allows for high device sensitivity and thermal isolation. The larger the separation distance, the higher the thermal resistivity between the membrane and the substrate, resulted in reduced parasitic heat transfer and increased thermal sensitivity. A further example of such an application is given in Figure 9.27.

Removal of the porous silicon creates free-standing structures for thermal isolation. Alternatively, the porous silicon can be left as porous silicon or oxidized to make a thick oxide. The thermal conductivity of silicon is of the order of 156 W/(m K). Porous silicon, on the other hand, has a thermal conductivity in the range of 0.1–2 W/(m K), depending on the porous structure, which can be improved slightly with oxidation [32].

The above structures used sacrificial porous silicon for thermal isolation. It can also be used for mechanical isolation to improve Q-factors for resonators. Figure 9.28 shows

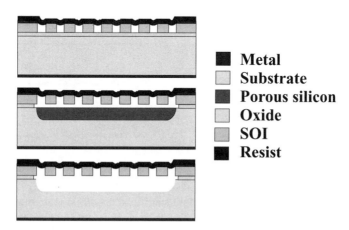

FIGURE 9.28. Process flow for a resonating structure with an increase in Q-factor using sacrificial porous silicon. Taken from [33].

FIGURE 9.29. An example of an accelerometer fabricated using selective sacrificial porous silicon [22].

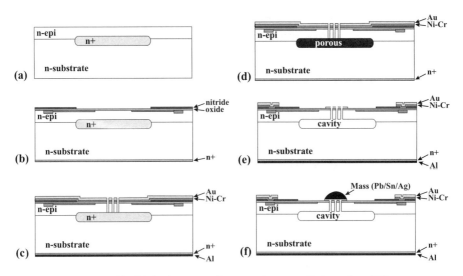

FIGURE 9.30. Fabrication process for an accelerometer. Redrawn from [33].

an example of a process flow [33]. Without sacrificial porous silicon, the final air gap would be limited to the oxide thickness. The porous silicon technique enables this gap to be significantly increased. In this case, air gap of 80 μm was achieved. An increase in air gap from 2 μm to 30 μm yielded an increase in Q-factor of 100%.

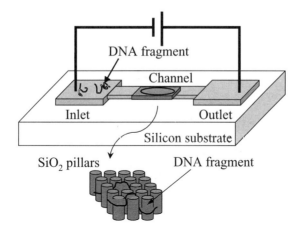

FIGURE 9.31. Basic structure of the DNA separation chip. Taken from [34].

9.3.4. Accelerometer

A number of processes have been proposed for the fabrication of accelerometers using both micro and macroporous silicon. A lateral accelerometer has been fabricated using an n-type epi as the mechanical material and the p-type substrate as the sacrificial layer [22]. An example of one of these structures is given in Figure 9.29. Macroporous silicon has also been applied, as shown in Figure 9.16.

A process using an n+ buried layer as a sacrificial layer is given in Figure 9.30 [34]. The relative etch rate of n+- and n-type silicon was given in Figure 9.2.

The mass is supported by eight cantilevers and the output through four Wheatstone bridges. Using an amplifier with amplification of 200, the accelerometer yields an output of approximately 0.24 V/g using all four bridges. The resonant frequency has been measured at 2.82 kHz.

9.3.5. DNA Separation Chip

Macroporous silicon can be used to make oxide pillars suitable for a DNA separation chip. The pillars need to be in a channel and an electric field is placed between the ends of the channel to pull the DNA through. The structure of this device is given in Figure 9.31 [35]. The basic process sequence is shown in Figure 9.32. Firstly, straight holes are etched using the macroporous silicon process described above (Figure 9.32a). Leaving the nitride mask in place the inside of the pores are oxidized (Figure 9.32b). The nitride mask is then removed and the silicon etched back using TMAH, which does not etch the oxide (Figure 9.32d). The advantage of this approach is that the spacing between the resulting pillars can be accurately controlled by the oxidation and reduced to well below the lithography resolution. The resulting pillar structure is given in Figure 9.33.

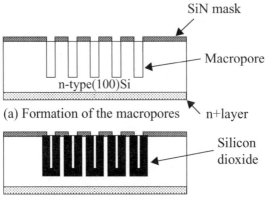

SiN mask

Macropore

n-type(100)Si

(a) Formation of the macropores   n+layer

Silicon
dioxide

(b) Thermal oxidation along the macropores

(c) Mechanical polishing for the wafer surface

(d) Fabrication of SiO$_2$ pillars by TMAH etching

FIGURE 9.32. DNA separation chip process.

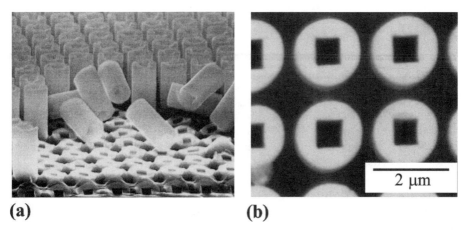

(a)   **(b)**

FIGURE 9.33. Pillar structure: (a) SEM photographs of pillars and (b) optical photograph illustrating the distance between the pillars.

9.4. CONCLUSIONS

Porous silicon was first observed during electropolishing when the current density was too low for a given HF concentration. Interest was evoked in the early 1990s with the discovery of photoluminescence and the potential for micromachining. Since these early investigations, both macroporous and microporous silicon have been shown to be valuable tools for micromachining.

REFERENCES

[1] M. Esashi, H. Komatsu, T. Matsuo, M. Takahashi, T. Takioshima, K. Imabayashi and H. Ozawa, Fabrication of catheter-tip and sidewall miniature pressure sensor, IEEE Trans. Electron Devices **29**, 57–63 (1982).

[2] C.J.M. Eijkel, J. Branebjerg, M. Elwenspoek and F.C.M. van de Pol, A new technology for micromachining of silicon dopant selective HF anodic etching for the realization of low-doped monocrystalline silicon structures, IEEE Electron Device Lett. **11**, 588–589 (1990).

[3] W. Lang, P. Steiner, A. Richter, K. Marusczyk, G. Weimann and H. Sandmaier, Application of porous silicon as a sacrificial layer, *Proceedings Transducers'93*, June 1993, Yokohama, Japan, pp. 202–205.

[4] P. Steiner and W. Lang, Micromachining applications of porous silicon, Thin Solid Films **255**, 52–58 (1995).

[5] V. Lehmann, Porous silicon—a new material for MEMS, *Proceedings MEMS'96*, February 1996, San Diego, pp. 1–6.

[6] F. Roozeboom, R. Elfrink, J. Verhoeven, J. van den Meerakker and F. Holthuysen, High-value MOS capacitor arrays in ultradeep trenches in silicon, Microelectron. Eng. **53**, 581–584 (2000).

[7] H. Ohji, P.J. Trimp and P.J. French, Fabrication of free standing structures using a single step electrochemical etching in hydrofluoric acid, Sensors Actuators A: Phys. **73**, 95–100 (1999).

[8] G.M. O'Halloran, M. Kuhl, P.J. Trimp and P.J. French, The effect of additives on the absorption properties of porous silicon, Sensors Actuators A **61**, 415–662 (1997).

[9] M. Kuhl, G.M. O'Halloran, P.T.J. Gennissen and P.J. French, Formation of porous silicon using an ammonium fluoride based electrolyte for application as a sacrificial layer, J. Micromech. Microeng. **8**, 317–322 (1998).

[10] H. Ohji, P.J. French and K. Tsutsumi, Fabrication of mechanical structures in p-type silicon using electrochemical etching, Sensors Actuators A: Phys. **82**(1–3) 254–258 (2000).

[11] R.L. Smith, S.-F. Chuang and S.D. Collins, Porous silicon morphologies and formation mechanism, Sensors Actuators **A21-A2** 825–829 (1990).

[12] S. Izuo, H. Ohji, K. Tsutsumi and P.J. French, Electrochemical etching for n-type silicon using a novel etchant, *Proceedings Transducers'01*, June 2001, Munich, Germany.

[13] L.T. Canham, Silicon quantum wire array fabrication by electrochemical dissolution of wafers, Appl. Phys. Lett. **57**, 1046–1048 (1990).

[14] H. Kaneko, P.J. French and R.F. Wolffenbuttel, Photo- and electro-luminescence from porous Si, J. Luminescence **57**, 101–104 (1993).

[15] Z.Y. Xu, M. Gal and M. Gross, Photoluminescence studies on porous silicon, Appl. Phys. Lett. **60**, 1375 (1992).

[16] P. Steiner, F. Kozlowski and W. Lang, Light-emitting porous silicon diode with an increased electroluminescence quantum efficiency, Appl. Phys. Lett. **62**, 2700–2702 (1993).

[17] H. Wong, Recent developments in silicon optoelectronic devices, Microelectron. Reliab. **42**, 317–326 (2002).

[18] G.M. O'Halloran, P.M. Sarro, J. Groeneweg, P.J. Trimp and P.J. French, A bulk micromachined humidity sensor based on porous silicon, *Proceedings Transducers'97*, 16–19 June, Chicago, USA, 1997, pp. 563–566.

[19] C. Baratto, G. Faglia, G. Sberveglieri, L. Boarino, A.M. Rossi and G. Amato, Front-side micromachined porous silicon nitrogen dioxide gas sensor, Thin Solid Films **391**, 261–264 (2001).

[20] V. Lysenko, S. Périchon, B. Remaki and D. Barbier, Thermal isolation in microsystems with porous silicon, Sensors Actuators A **99**, 13–24 (2002).

[21] T.E. Bell, P.T.J. Gennissen and M. Kuhl, Porous silicon as a sacrificial material, J. Micromech. Microeng. **6**, 361–369 (1996).

[22] P.T.J. Gennissen, P.J. French, D.P.A. de Munter, T.E. Bell, H. Kaneko and P.M. Sarro, Porous silicon micromachining techniques for acceleration fabrication, *Proceeding ESSDERC'95*, 25–27 September 1995, Den Haag, The Netherlands, pp. 593–596.

[23] P.T.J. Gennissen, H. Ohji, P.J. French, C.M.A. Ashruf, G.M. O'Halloran and P.M. Sarro, Combination of epipoly and electropolishing for fabrication of accelerometers with large substrate separation gaps, *Proceedings Eurosensors'99*, 13–15 September 1999, Den Haag, The Netherlands, pp. 1029–1032 (CD-ROM version).

[24] H. Ohji, P.J. French, S. Izuo and K. Tsutsumi, Initial pits for electrochemical etching in hydrofluoric acid, Sensors Actuators A: Phys. **85**(1–3), 390–394 (2000).

[25] E.K. Propst and P.A. Kohl, The electrochemical oxidation of silicon and formation of porous silicon in acetonitrile, J. Electrochem. Soc **141**, 1006–1013 (1994).

[26] E.A. Ponomarev and C. Levy-Clement, Macroporous formation on p-type Si in fluoride containing organic electrolytes, Electrochem. Solid-State Lett. **1**, 42–45 (1998).

[27] R.C. Anderson, R.S. Muller and C.W. Tobias, Porous polycrystalline silicon: a new material for MEMS, J. MEMS **3**, 10–17 (1994).

[28] G. Lammel and Ph. Renaud, Free-standing, mobile 3D porous silicon microstructures, Sensors Actuators A: Phys. **85**(1–3), 356–360 (2000).

[29] W. Lang, P. Steiner, U. Schaber and A. Richter, A thin film bolometer using porous silicon technology, Sensors Actuators A **43**, 185–187 (1994).

[30] F. Hedrich, S. Billat and W. Lang, Structuring of membrane sensors using sacrificial porous silicon, Sensors Actuators A **84**, 315–323 (2000).

[31] Cs. Dücsö, É. Vázsonyi, M. Ádám, I. Szabó, I. Bársony, J.G.E. Gardeniers and A. Van den Berg, Porous silicon bulk micromachining for thermally isolated membrane formation, Sensors Actuators A **60**, 235–239 (1997).

[32] V. Lysenko, S. Périchon, B. Remaki and D. Barbier, Thermal isolation in microsystems with porous silicon, Sensors Actuators A **99**, 13–24 (2002).

[33] H. Artmann and W. Frey, Porous silicon technique for realization of surface micromachined silicon structures with large gaps, Sensors Actuators A: Phys. **74**(1–3), 104–108 (1999).

[34] J.-H. Sim, Ju.-H. Lee, Jo-H. Lee, C.-S. Cho and J.-S. Kim, Eight-beam piezoresistive accelerometer fabricated by using a selective poroussilicon etching method, Sensors Actuators A **66**, 273–278 (1998).

[35] H. Ohji, S. Izuo, P.J. French and K. Tsutsumi, Pillar structures with a sub-micron space fabricated by macroporous-based micromachining, Sensors Actuators A **97–98**, 744–748 (2002).

Index